THE COMMISSION OF THE EUROPEAN COMMUNITIES

Ocean Disposal of Radioactive Waste by Penetrator Emplacement

This report was prepared for
the European Atomic Energy Community's
Cost-sharing Research Programme on
"Radioactive Waste Management and Disposal"
Under contract No 394-83-7 WAS UK
Project 7B: "Disposal in Sea-bed Geological Formations"

Ocean Disposal of Radioactive Waste by Penetrator Emplacement

OVE ARUP & PARTNERS
13 Fitzroy Street, London W1P 6BQ

Published by GRAHAM & TROTMAN Ltd
for the Commission of the European Communities

Published in 1985 by Graham & Trotman Ltd
66 Wilton Road, London SW1V 1DE, UK.
for the Commission of the European Communities,
Directorate-General Information Market and Innovation,
Luxembourg

EUR 10170 EN

ISBN 0 86010 812 0

Printed and bound in Great Britain

CONTENTS

SUMMARY

A number of options for the disposal of vitrified high level radioactive waste are being studied to ensure that safe methods are available when the time comes for disposal operations to commence. This study has considered the engineering and operational aspects of the Penetrator Option for ocean disposal to enable technical comparisons with other options to be made.

In the Penetrator Option concept, waste would be loaded into carefully designed containers which would be launched at a suitable deep ocean site where they would fall freely through the water and would embed themselves completely within the seabed sediments. Radiological protection would be provided by a multi-barrier system including the vitrified waste form, the penetrator containment, the covering sediment and the ocean.

Calculations and demonstration have shown that penetrators could easily achieve embedment depths in excess of 30m and preliminary radiological assessments indicate that 30m of intact sediment would be an effective barrier for radionuclide isolation. It may also be concluded from theoretical studies that the entry holes made by penetrators would close but this has yet to be demonstrated.

The integrity of the sediment barrier may be affected by the temperature rise following disposal and further research of sediment thermal response is required. The peak sediment temperature for waste disposed of after 50 years storage is calculated to be 170°C, whereas after 100 years storage about 60°C is predicted. The possibility of limiting peak sediment temperatures by extended storage should be considered for ocean disposal concepts involving heat generating waste.

The study concludes that a 75mm thickness of low carbon steel appears to be sufficient to provide a containment life of 500 to 1000 years during which time the waste heat output would have decayed to an insignificant level. However further corrosion research is needed to confirm the choice of metal thickness.

A reference penetrator design has been produced and has been used to study possible disposal operations. The penetrator manufacture, assembly and waste disposal operations are all considered to be feasible using present day technology although demonstration is still required.

The area of seabed required for a disposal rate of 85m^3 per year has been calculated to be 1.9 km^2 per year. This rate of usage is consistent with the capacity of the Great Meteor East study area which is being assessed against waste emplacement suitability criteria by an expert group of the OECD Nuclear Energy Agency.

Disposal costs have been assessed with the primary objective of identifying major cost items and cost sensitivity. The cost of disposal has been calculated at 1984 prices to be approximately 0.007p per kW hr of electricity generated.

FIGURES

1.0 INTRODUCTION

1.1 Ocean Disposal of High Level Waste

High Level Waste (HLW) in the form of a highly radioactive and heat generating liquor is produced when spent nuclear fuel is reprocessed to recover uranium and plutonium. The HLW is solidified in a subsequent vitrification process.

Emplacement in stable geological formations is currently seen as a possible way of isolating the HLW for its radioactivity to decay below levels which would pose an unacceptable health risk to man or pollution hazard to the environment.

The radioactive waste management programme supported by the UK Department of the Environment (DoE) and the Commission of the European Communities (CEC) is considering repositories located in continental geological formations, on the deep ocean floor and beneath the deep ocean floor (Ref. 1).

Whilst this study considers only disposal beneath the deep ocean floor it takes due account of the research into other repository locations.

1.2 Penetrator Option

Conceptual studies (Refs. 2 to 7) have enabled two basic methods of emplacing HLW beneath the deep ocean floor to be identified for detailed study. Other ocean disposal methods have not been discounted but are not being actively investigated at this stage.

The first method is to drill holes in the seabed, lower in strings of waste canisters and to seal the hole.

The second method is to drop free fall projectiles loaded with waste at sites where they will completely embed themselves in the seabed sediments. This is known as the penetrator option.

The possible use of penetrators for HLW emplacement presents geotechnical problems for which in situ data are very sparse and existing theory will need development.

A little work has been done (Refs. 8 and 9) to estimate the strengths of sediments in relatively shallow water using calibrated penetrometers and this was assessed in previous penetrator emplacement studies (Ref. 7).

Current work is directed at developing theoretical models for the penetration process and validating the theories produced by laboratory experiments and field trials.

1.3 Objectives

The objectives of this study were to consider the engineering and cost implications of the penetrator option for the deep ocean disposal of HLW and to attempt to establish the technical and operational feasibility of the method to enable comparisons to be made with alternative options.

The report comprises an introductory section which sets the penetrator option in the general perspective of HLW disposal research, Part 1 which describes engineering aspects of a wide variety of penetrator design concepts and Part 2 which describes manufacturing and operational techniques for a reference penetrator design.

Additional data and information in support of the discussions in the text are contained in appendices at the end of the report.

The Penetrator Engineering Study was carried out in two phases. This report describes the work done in both phases and includes, as Part 1, a summary of the report which was produced on completion of Phase 1 (Ref. 10). Chapter 8 of this report was considerably influenced by the results of a penetrator trajectory modelling study (Ref. 21) carried out in parallel with this study but under a separate contract.

All the work described above was carried out by Ove Arup and Partners as part of a coordinated programme of research including desk studies, laboratory experiments and deep ocean trials funded by the UK Department of the Environment and the Commission of the European communities.

The help and assistance of Mr. T.J. Freeman and other members of the Building Research Establishment staff is acknowledged for managing the work on behalf of the UK Department of the Environment.

1.4 Site Selection

The success of the penetrator option for the disposal of HLW in the deep ocean depends on the existence of suitable sites.

For the purposes of this study, three areas have been selected from those being considered in detail by the Site Selection Task Group (SSTG) of the Nuclear Energy Agency Seabed Working Group (SWG). Oceanographic surveys have been carried out to find the thickness, area, stratigraphy and other relevant physical parameters of each study area.

These areas are shown on Figure 1 and the relevant data are given in Appendix A.

Detailed site investigation information is reported by SSTG in the proceedings of the annual meetings of the SWG.

1.5 Engineering Studies

Engineering studies (Ref. 7) predicted that an embedment depth of 30m was achievable for a wide range of possible penetrator designs and that the entry hole would close behind the penetrator. The encouraging nature of these predictions was qualified, however, by the lack of reliable in situ geotechnical data for deep ocean sediments.

Confidence in the feasibility of the option has been increased following the success of a series of deep ocean trials carried out in March 1983 when embedment depths of more than 30m were recorded for two tonne model penetrators (Ref. 11).

The international status of emplacement techniques research and study at the end of 1982 is reported in Reference 12.

1.6 Radiological Assessment

The primary objective of any HLW disposal system will be to minimise the radiological impact on the environment. The engineering criteria to be developed must therefore be those most likely to lead to a favourable outcome when the detailed penetrator design is assessed in radiological terms.

The International Commission on Radiological Protection (ICRP) optimisation principle, which states that all exposures shall be kept as low as reasonably achievable (ALARA), economic and social factors being taken into account, has been adopted as the governing radiological protection principle for this study.

Optimisation of radiological protection is in practice an onerous requirement as it entails understanding how the radiological protection system is affected by a wide range of parameters.

1.7 Radiological Barriers

Radiological assessments of the penetrator option will consider a number of barriers which prevent or delay the migration of radionuclides from the HLW to man. In this study four major barriers have been identified.

The first barrier is the waste form which will be a vitrified solid block in a canister. The thin metal canister itself is not considered a barrier.

The second barrier is the sealed body shell of the penetrator or an alternative sealed containment surrounding the waste canister. This is called the engineered barrier although the word overpack is sometimes used to describe a containment vessel for a waste canister.

The third barrier is the seabed sediment between the emplaced waste and the ocean floor and the fourth is the ocean itself.

The waste form is highly insoluble and minimises leaching of radionuclides from the glass matrix. The leach rate is, however, highest during the first few hundred years after reprocessing, as the fission products decay heat generated in this period significantly increases the glass solubility. Pore water flow around the emplaced waste may also be increased by the waste heat output.

The engineered barrier completely prevents the ingress of water and offers a total containment during its design life. Unfortunately some of the radionuclides have half-lives of thousands of years or more and it is unlikely that any engineered barrier could be specified to have a sufficiently long design life to completely contain these indefinitely.

The sediment barrier ideally has a very low permeability to reduce pore water flows to the minimum. The sorptive properties, which retard the migration of nuclides by a range of chemical and physico-chemical interactions such as ion exchange, precipitation and absorption, are also important. The sediments must have geotechnical properties which will allow a penetrator to embed itself and which will permit healing so that the entry hole closes and disturbance is minimised. The design life of the engineered barrier provides a period of time for sediment barrier healing to occur.

The ocean barrier acts by diluting the eventual release from the sediments and the low circulation rate of water near the seabed delays dispersion allowing a further period of time for decay thereby reducing the average dose levels to man.

1.8 Research Requirements

This engineering study uses the results and conclusions of initial generic studies in order to assess the feasibility of the penetrator option.

Research is still in progress in many areas and, as yet, the generic studies have not encompassed every aspect of the penetrator method necessary to demonstrate feasibility.

In particular, further research is required to demonstrate whether or not sediment entry holes close behind penetrators and to define the sediment properties needed for the predictive models which will be used to assess the safety of the disposal method.

These areas of uncertainty dictate that any assessment of feasibility will necessarily be conditional and it will be necessary to carry out further research before feasibility can be finally confirmed.

1.9 Basic Data

The basic data for HLW characteristics and deep ocean study areas which have been used in this study are set out in Appendix A.

PART 1 GENERAL ASSESSMENT OF OPTIONS

2.0 PENETRATOR DESIGNS

2.1 Introduction

This chapter deals with the performance requirements for penetrators and then discusses candidate materials for manufacture before proposing penetrator designs.

There are three fundamental performance requirements for penetrators as well as several secondary but important criteria. The fundamental requirements are:-

a) to be compatibile with the waste form
b) to provide an engineered barrier
c) to achieve adequate embedment performance

The basic shapes and sizes of the penetrators are designed to receive the HLW canisters which will be used in the UK and France with a minimum of modifications. In practice it is found that the requirement to provide an engineered barrier exerts most influence on penetrator design because initial research indicates that satisfactory embedment can be achieved by a wide range of penetrator sizes and weights.

Secondary factors, such as ease of fabrication and transport, for the penetrator design options are also discussed.

2.2 Main Assumptions

The main initial assumptions of this study are detailed below to assist in the understanding of the course chosen and methodology adopted for the development of penetrator designs.

2.2.1 Feasibility criteria

In the UK vitrified waste will be stored above ground for at least 50 years before disposal (Ref. 13).

As it is difficult to forecast engineering practices in 50 years time, existing technology has been adopted wherever possible to assess technical and operational feasibility.

As a consequence, the feasibility criteria applied to penetrator designs developed in this study are based on:-

a) existing technology including sophisticated techniques actively under development for industrial application, or

b) existing technology adapted for specific application to penetrators providing that the practicality of the proposed modification can be judged with a high degree of confidence

2.2.2 Penetrator size limits

An important consequence of the adopted feasibility criteria is that existing industrial practices can be used to help identify suitable maximum sizes for penetrators with regard to manufacturing and handling capacities. At this stage it has been considered that penetrators in excess of 20m in length would present unreasonable handling difficulties. Therefore a range of penetrator designs to contain between one and twelve HLW canisters has been considered.

2.2.3 Disposal operation scale

The number of penetrators required per annum is determined by their capacity and the rate of waste disposal, data for which is given in Appendix A. Assuming the largest capacity and the lowest disposal rate then 50 penetrators per annum would be required. However, if the smallest capacity and largest disposal rate are assumed then 2670 penetrators per annum would be required.

It is clear that for these numbers of penetrators simple designs suitable for production line construction techniques would be preferred and that excessive design sophistication should be avoided if it leads to additional complications.

2.2.4 Unpowered penetrators

Free fall, fixed geometry penetrators could attain sufficient terminal velocity to achieve adequate embedment and could be designed to be stable and to follow predictable trajectories. The unnecessary complications of powered propulsion and steering have not been included in any of the designs proposed.

2.3 Waste Form Compatibility Requirement

As stated in 2.1, the first of the three primary performance requirements is quite simply that penetrators should contain HLW in the form that it is produced by vitrification plants.

2.3.1 HLW vitrification

The most widely adopted concept is to cast molten vitrified HLW in metal canisters.

The canisters are filled to a nominal 80% of their capacity leaving an ullage so as to avoid the possibility of overflowing and spattering when the HLW is poured in. In some circumstances it has been noted that voids can be trapped in HLW as it solidifies in a canister (Ref. 14). This could seriously reduce the ability of the penetrator contents to resist external pressure and is discussed further in 2.7.8.

As well as acting as moulds the canisters prevent the disintegration of the glass-like HLW into unmanageable fragments and facilitate mechanical handling during production and storage.

2.3.2 Assumed canister design

The canister form assumed in this study for penetrator designs is shown in Figure 2. It is based on designs which has been selected for both the French and British vitrification programmes.

The head of the canister is pintle shaped so that it can be readily held in a mechanical grab and the champagne bottle bottom shape of the base allows canisters to be stacked vertically. The lid is provided to prevent HLW fragments escaping and is welded in position.

2.3.3 Alternative canister sizes

Other canister sizes have been considered in various vitrification research programmes. Invariably cylindrical, the main proposals have ranged roughly from 0.25 to 0.50m in diameter and from 1.0 to 3.0m in height.

This report contains no detailed discussion of alternative canisters because the technical and operational feasibility of HLW disposal penetrators is not considered to be sensitive to canister sizes within the approximate range given above.

2.3.4 Canister material

Materials with suitable high temperature properties are required for the canisters. Various materials such as stainless steels and nickel alloys like Inconel are used or proposed for canisters designed by different organisations.

The canister material employed has little significance for the design of penetrators. In general, therefore, no particular canister material requirements for disposal by penetrators have been identified.

2.3.5 Alternative HLW forms

Alternatives to HLW vitrification have been proposed at various times and their future adoption has not been ruled out. Examples include solidification of HLW into corundum or Synroc (synthetic rocks) or ceramic forms. There is no obvious reason why penetrators should not be considered for ocean disposal of waste in these forms.

2.3.6 Spent fuel disposal

Currently, some countries do not consider fuel reprocessing to be economically justified and are investigating long term storage and disposal for spent fuel rods from power stations.

The geometry of spent fuel bundles would allow them to be fitted into penetrators but their complex shape would mean that greater attention would need to be paid to void filling and various other details. It is also important to note that ocean disposal options are unlikely to permit fuel retrieval for subsequent reprocessing should this become desirable.

The use of penetrators for spent fuel disposal has not been considered in this study.

2.3.7 HLW heat and radiation outputs

The relationships assumed for the heat and radiation generation decay with age of vitrified HLW are given in Figures 3 and 4. The actual heat and radiation outputs will depend on the exact composition of waste.

The curves clearly show the significance of providing several decades of interim storage and 500 to 1000 years containment using an engineered barrier. It should also be noted that even 300mm thick steel shielding would be insufficient to allow prolonged periods of "hands on" operations during penetrator assembly or disposal.

2.3.8 Penetrator design ramifications

The ramifications for penetrator design of the adoption of the Figure 2 canister size are the limitations placed on the selection of penetrator body internal diameter and length increments. The circular cross-section of the canisters makes the selection of similar circular cross-section penetrator designs most appropriate.

2.4 Engineered Barrier Requirement

Consideration of means of incorporating the engineered barrier discussed in 1.7 has been found to have most influence on the development of penetrator designs.

2.4.1 Overpacks

In studies of continental geological disposal options, a container, usually made of metal, acting as an engineered barrier for the HLW is sometimes described as an overpack.

2.4.2 Engineered barrier options

Two ways of providing an engineered barrier have been considered in this study.

The first way is by the use of overpacks which can either be loaded into a separate body shell which need not itself offer any containment or can be joined together to form a penetrator with only the addition of nose and tail units. These options are shown in Figures 5 and 6 respectively.

The second way is by designing the body shell to act as a containment. This integrated design approach is used for the penetrators shown in Figures 7 and 8 and is currently considered to be the most appropriate construction.

2.4.3 HLW canisters

It is not considered feasible for canisters to be designed to act as moulds for molten HLW and to provide adequate resistance to corrosion in the disposal environment. The canister is therefore ignored in estimates of the containment lives of engineered barriers.

2.4.4 Containment design life

The period of isolation that the engineered barrier is required to provide is called containment design life. The life of the engineered barrier will be less than the expected duration of the other isolating barriers, such as the seabed sediments, by many orders of magnitude. Preliminary radiological assessments (Ref. 15) have considered various periods of total containment corresponding to the life of the engineered barrier. The results have been shown to be relatively insensitive even for zero years containment.

The current lack of detailed radionuclide migration data means that radiological assessment models cannot as yet be refined sufficiently to identify an optimum containment design life. Therefore, a design life has had to be selected using engineering judgement taking into account the many diverse factors involved.

Research on HLW disposal in continental geological formations has led to the consideration of 500-1000 years as the required life for containment. As discussed in 2.3.7, the timescale recognises that after 300 years or so the heat output will have reduced to insignificant levels.

Preliminary radiological assessments of the penetrator option have suggested that there is little benefit in extending containment life beyond 500 years unless taken beyond 4000 years. It is not considered practical to confirm the feasibility of this because of the difficulties of assessing such very long term corrosion behaviour of materials. Thus extremely long design lives, such as 4,000 years, cannot be predicted with confidence.

A high degree of confidence is required when dealing with such highly toxic materials as HLW and a conservative approach is justified. Many assumptions made regarding the penetrator option will be extremely difficult to verify. These are both arguments in favour of providing a long term containment.

There are a number of other benefits attributable to a relatively long containment design life of, say, 500 years.

These include:-

a) higher probability of a more favourable detailed radiological assessment, including improvements for most accident situations

b) time for any thermally induced sediment disturbance to heal after heat output has become insignificant

c) time for hole closure to become effective including healing after heat induced disturbance

d) time for short lived radionuclides to decay

e) increased embedment depth due to the additional weight of a long design life containment

f) increased public confidence

A zero year design life containment implies that full confidence must be placed in the other waste isolation barriers and that they must become effective very soon after emplacement.

Apart from the reduction in weight and the greater ease of manufacture, there appear to be few advantages in performance of penetrators with no containment design life. Clearly a zero year design life leads to a more economical design, but confidence in its effectiveness is lower and so has been discounted for the purposes of this study.

Consideration of all these points suggests that the provision of a long term containment is a reasonable assumption at this stage and so a 500 year design life requirement has been adopted for this study.

This may subsequently be seen as too conservative but it has the advantage that the range of problems considered should simplify any future assessment of shorter design lives. Should further trials and research demonstrate that satisfactory embedment depths and hole closure would be readily achieved then this conservative approach could be reviewed.

It is also noted that the criterion for containment life specification is currently very severe. Failure is considered to have occurred, and full scale leaching of HLW to have commenced, as soon as any breach is created in the engineered barrier. In practice, the first breach would develop at an area of localised corrosion and by its very nature only affect a small proportion of the whole containment. The remaining containment material would clearly restrict the access of pore water for HLW leaching until complete degradation by general corrosion processes occurred. Thus better developed models of barrier effectiveness could lead to the acceptance of a less demanding design life requirement given in terms of both local and general degradation times.

2.4.5 Disposal environment

There is currently a lack of data for the seabed environment at the possible penetrator embedment depths although information on likely sea and pore water compositions can now be compiled (Ref. 16) and the seabed sediments are being investigated at several locations by members of the SWG (Ref. 12).

Even when more detailed information becomes available it is recognised that it will remain a very demanding task to predict accurately the influence of HLW disposal on the seabed environment.

First the seabed formation above and immediately adjacent to a penetrator will have been disrupted during the embedment event. Thereafter, radiation emitted by the HLW will have an ionising effect on surrounding water. The dissipation of heat generated by the HLW could cause pore water movements and mineralogical changes and possibly produce fissures in the proximate sediments (Ref. 17). The limitation of peak sediment temperature rises is discussed in 8.0. The influence of high hydrostatic pressure, the interaction between the seabed and corrosion products as the materials used in the penetrators degrade and the effects of any biological activity stimulated by the HLW radiation and heat outputs also need to be considered.

Characterisation of the disposal environment is, therefore, very complex and not fully evolved. This limits the accuracy and confidence with which corrosion mechanisms and rates can be developed for penetrator materials.

The current paucity of disposal environment information means that only qualitative remarks can be made about many important features. Confirmation or refinement of penetrator design proposals will only be possible as the ability to model the disposal environment is improved. A good example is the need for detailed assessment of likely pore water movements around a penetrator, to give a means of estimating the supply of dissolved oxygen and oxygen released by ionizing radiation as this is the limiting factor on corrosion rates for several important proposed penetrator materials.

2.4.6 Hydrostatic pressures

The water depths at the disposal study areas under consideration have led to the adoption of $60N/mm^2$ as a nominal hydrostatic pressure for penetrator design. It should be noted that this is significantly greater than the critical pressure, about $22N/mm^2$, of water above which there is no discernible difference between the liquid and gas phases or, crudely, the point when boiling does not occur because steam does not separate from water.

The choice of $60N/mm^2$ design pressure includes sufficient allowance for the additional stresses due to the sediments around embedded penetrators.

2.4.7 Penetrator design ramifications

The main ramification for penetrator design of meeting the engineered barrier requirement is the selection of appropriate combinations of material types and thicknesses. The assessment of likely corrosion rates and the behaviour of proposed penetrator designs under hydrostatic pressure are reviewed in 2.7.

2.5 Embedment Requirement

The requirement that penetrators would have to achieve satisfactory embedments at selected suitable seabed disposal sites is now reviewed. The method used to predict embedment depths is described in Appendix B.

2.5.1 Embedment depth

In the multiple barrier concept it is intended that the seabed sediments around buried HLW would provide the longest and thus most important period of isolation from man's environment. Preliminary radiological assessments considered a 30m burial depth as a reference value in the absence of definite penetrator design and performance information. A recent study by Wuschke et al (Ref. 18), as part of a probabilistic assessment demonstration project, concluded that there would be little advantage of burial at greater than 30m depths with regard to radiological protection. However, current radionuclide migration models do not reveal any minimum embedment at which the radiological consequences would exceed the limits recommended by the ICRP (Ref. 19).

For this study a requirement to achieve a minimum tail embedment of 10m has been identified as a reasonable assumption at this stage of penetrator option investigations. The relatively small increase in the radiological consequences predicted by current preliminary models for 10m as opposed to 30m burial has been demonstrated by Hill and Delow (Ref. 15).

The embedment depths calculated for many of the penetrator designs assessed in this study and displayed in Figures 9-12 are significantly greater than the minimum requirement of 10m discussed above. Generally embedments increase as a result of increasing penetrator length and weight. It has been assumed that the attainment under normal conditions of these deeper embedments would increase confidence in the suitability of the penetrator option for HLW ocean disposal. For example, it would be possible to demonstrate that a wide range of adverse conditions such as an undetected region of unusually stiff seabed sediments at a disposal site, would not reduce penetrator embedments below satisfactory depths.

Conversely, it is also arguable that very deep embedment would not necessarily be seen as advantageous because there is less detailed information available on the sediment properties at the greater depths.

However, the main conclusions of this engineering study would not be affected if a maximum penetration were established because designs could be adjusted to limit expected embedment to pre-selected depths which are less than the maximum attainable by a fully streamlined penetrator of the same weight.

2.5.2 Embedment path

Any tendency for a penetrator to follow a significantly curving or "J" path through the seabed would be detrimental because it would lessen the final depth below the seafloor that could be achieved.

However, there is nothing to suggest that the penetrator designs, especially those with length to diameter ratios exceeding about eight, evolved in this study would be prone to follow curved embedment paths. The relatively large and robust fins provided for stability during free-fall descent through the ocean would also be expected to have a controlling effect on embedment into the seabed. Support for these arguments is provided by the conclusions of research programmes in the USA (Ref. 20) and the UK (Ref. 21).

Therefore, no particular restraints on penetrator designs have been adopted with respect to embedment path control though this will be subject to the satisfactory outcome of future trials.

2.5.3 Hole Closure

The radiological protection afforded by the sediment barrier cannot be guaranteed if the penetration hole does not close. To a lesser extent the total degree of disturbance caused by emplacement must be quantified to assess fully the sediment barrier competence.

The technical problems of describing and assessing the mechanisms of hole closure are new and, as yet, unresolved. Some researchers have considered the problem by assessing the prevailing water and soil pressures (Refs. 22, 23 and 24) and preliminary findings indicate that the entry holes will close although this awaits experimental verification.

It is considered likely that the embedment path hole will close behind a penetrator if only because of the very soft, almost fluid properties of the upper sediment layers. At this stage, therefore, no penetrator design features have been identified as necessary to ensure hole closure.

2.5.4 Seabed impact

A basic design requirement is that penetrators should not be susceptible to damage when they hit and then penetrate the seabed. In practice it does not seem that this will be an arduous requirement to satisfy despite the high impact velocities of up to 100m/s for the largest penetrators. Preliminary calculations and the results of trials (Ref. 11) indicate that impact and penetration decelerations would be modest; probably being less than ten times the acceleration of gravity.

2.5.5 Sediment characteristics

As suggested previously, the main characteristic affecting penetrator embedment performance at a seabed disposal site is the low strength of the sediments. This is particularly true of the surface layers which have almost zero strength. The undrained shear strength, c_u, of the sediments is the index property which mostly affects the calculations of probable embedment depths for penetrators.

The design profile for c_u assumed for calculations of penetrator embedment depths is discussed in Appendix B.

Information regarding seabed sediments is published by the SWG in the annual reports of its seabed sampling and analysis programme.

2.5.6 Penetrator design ramifications

The constraints on penetrator design in order to achieve satisfactory embedment performance are considered to be the adoption of smooth streamlined profiles and the assumption that fins provided mainly to stabilise the free-fall descent through the ocean will also influence the course of seabed embedment. Fins should thus be sufficiently robust to survive loads induced during seabed embedment. Attainable embedment depths are enhanced as a function of increasing penetrator length and weight.

2.6 Secondary Performance Requirements

The classification of the following requirements as secondary is not intended to suggest that they are in any way optional specifications for penetrator performance. Indeed their satisfaction would be just as crucial to the acceptability of a penetrator design as meeting the previously discussed primary requirements. However, they are treated separately because they have been found to pose fewer constraints on penetrator details. Thus the penetrator designs to meet the three primary requirements discussed in 2.3, 2.4 and 2.5 are often also capable of meeting the secondary requirements.

2.6.1 Component fabrication

The most straightforward features compatible with satisfactory performance should be adopted to minimise component fabrication difficulties. Unnecessarily severe tolerance and material defect acceptance levels have been avoided for similar reasons.

Options requiring relatively conventional and commonly employed engineering materials and fabrication processes have been favoured because experience of their general use will avoid potential production and quality problems.

One advisable design criterion pertinent to component fabrication would be to select features so that all critical sections, which might include welds, could be checked adequately by Non-Destructive Examination (NDE) techniques. This requirement could entail, for example, the specification of smoother surface finishes to suit NDE than would otherwise be necessary for satisfactory penetrator performance.

The capability of manufacturers to fabricate components in the sizes and numbers necessary for a penetrator has to be assessed. It is not appropriate to allow this consideration to constrain the size of penetrator designs because of the impracticality of anticipating the manufacturing equipment and techniques that would be available in 50 or more years. However, the capacity of current steelworking processes has been evaluated to demonstrate this aspect of technical feasibility for the penetrator design options selected for further study.

2.6.2 Component transport

The transport of components from the fabrication works to a penetrator assembly plant by rail or public road has been considered for each penetrator design option. This essentially suggests maximum individual component lengths of about 12m and weights of about 30t to avoid the need for special transport arrangements. Penetrators with capacities for more than 5 to 8 HLW canisters depending on the design option would exceed these limits.

Components for large penetrators might have to be transported in sub-lengths for joining into finished lengths as an additional operation at the assembly plant.

2.6.3 Penetrator assembly

Consideration of the unusual aspects of penetrator assembly points to the advantages of penetrators with straightforward design features and carefully chosen realistic quality targets for similar reasons to those discussed in 2.6.1. In the context of this report, penetrator assembly refers to the loading of HLW canisters and all subsequent operations to complete a penetrator. These operations would be undertaken in shielded work areas at a specialised assembly plant because of radiation emission from the HLW. Component fabrication refers to the making of penetrator bodies, noses and other parts in conventional works for delivery to the assembly plant.

Most operations in the assembly plant would have to be remotely controlled which could restrict the range of possible procedures and would certainly favour relatively straightforward techniques.

Once assembled there would be special difficulties and perhaps a small risk increase if a penetrator had subsequently to be dismantled and its HLW contents transferred because an assembly defect exceeded acceptance standards. There would, therefore, be quantifiable disadvantages in setting unnecessarily severe defect acceptance levels or selecting a design of penetrator that was more sensitive to assembly defects than an alternative acceptable design.

2.6.4 Void filling

Any requirement to fill voids within HLW canisters and between canisters and penetrator bodies would be a part of penetrator assembly but is evaluated separately because of its significance for design development.

Void filling with structural materials could be necessary for one or more of the following reasons depending on the penetrator design option:-

a) to transmit pressures from penetrator bodies to their HLW contents.
b) to minimise water circulation on eventual failure of the containment
c) to hold canisters in position within penetrator bodies

It is almost certain that some or all voids would have to be filled and adequate internal clearances to allow the satisfactory injection of the filler materials must be provided.

The choice of filler material is an important subject discussed further in 2.7.8.

2.6.5 Penetrator transport

Options for transporting assembled penetrators on land and in disposal ships are discussed in 3.0 where the ramifications for assembly plant location are also assessed.

The maximum gross weight proposed for transport flasks from the point of view of present day handling and transport limitations is in the region of 120t. Penetrator options which would exceed this weight if carried in a transport flask designed to current IAEA regulations are noted in 2.8.8.

2.6.6 Penetrator storage

It has been assumed that batches of penetrators would be assembled and then stored until they could be taken aboard a disposal ship. Cooling, probably by air convection as discussed in 3.0 would be necessary because of the HLW heat output. However, temperature rises during storage would not be expected to approach the peak temperature limits following seabed disposal as discussed in 2.6.10. Therefore, no penetrator design features have been adopted specifically to meet the storage requirement.

2.6.7 Penetrator handling

It is possible to envisage many ways by which components and assembled penetrators could be held for lifting and other handling operations. They fall into two categories; those which require special lugs or recesses to be built into penetrator components and those which do not. At this early stage it is inappropriate to develop detailed handling procedures beyond that necessary to demonstrate technical feasibility. Therefore, no built-in handling features have been incorporated in the penetrator design proposals.

2.6.8. Pre-disposal contingencies

Pre-disposal contingencies is used to describe all possible unintentional incidents that could occur from the time penetrators are loaded with HLW canisters until they are embedded as planned at the disposal site. There are two main types of situation that could arise.

a) Incidents during assembly, storage, transport or handling on land or aboard a ship after which a penetrator could be retrieved for any necessary remedial work

b) Incidents involving loss at sea after a shipping accident or unsatisfactory embedment at a disposal site when it might be difficult to retrieve a penetrator.

A full assessment of credible incidents and their likely consequences would be required at the detailed design stage before a disposal operation could be approved by a licensing authority but is not required for the current level of study.

A preliminary evaluation of the incidents that would have to be considered does, however, lead to the conclusion that no special contingency criteria need to be applied during this initial development of penetrator design options for the following reasons.

a) the robustness of penetrator designs to resist seabed embedment loads and hydrostatic pressures will also be sufficient to survive most postulated mishandling and accident incidents during assembly and transport. Further aspects of this are discussed in 2.6.14.

b) the engineered barrier would mitigate the radiological consequences of accidental loss at sea.

2.6.9 Hydrodynamic speed and trajectory

A requirement for streamlined profiles to reduce hydrodynamic drag has been applied during penetrator design development. This maximises the terminal velocities that would be reached during free-fall descent through the ocean so enhancing attainable embedment depths in the seabed. Terminal velocities between about 25 and 100m/s for the penetrator design options assessed in 2.7 have been estimated using the hydrodynamic model described in Appendix B and are given on Figures 9 to 12.

The requirement for penetrators to take up near-vertical and stable trajectories during free-fall ocean descent is satisfied by the provision of tail fins. The necessary sizes of fins and the hydrodynamic loads on them are discussed in Appendix B.

The possible benefits of designing fins to induce spinning and spiralling of penetrators during the free-fall ocean descent are considered in 8.11.

2.6.10 Peak sediment temperature rise

The seabed sediments have a thermal conductivity which is very similar to the conductivity of the glass-like HLW but is low compared with that of metals. Therefore, once embedded in the seabed, the temperature rises until the rate of heat dissipation through the insulating sediments equals the rate of heat generation by the HLW.

A general principle for HLW disposal is that temperatures should not rise above specified ceilings so as not to impair the effectiveness of the radiological isolation afforded by the surrounding media. For deep ocean sediment a 100° C limit is commonly assumed pending the results of more detailed and site specific research.

Peak sediment temperature rise limitation is not treated as a primary penetrator design requirement because it is best effected by controlling the quantity, age and spacing of the waste at disposal.

However, it is worth noting that for a given waste canister diameter longer and smaller diameter penetrators cause more pronounced temperature effects and are thus at a disadvantage to shorter thicker walled options.

The interaction of heat fields around adjacent penetrators in a seabed disposal site is discussed in 8.5.1 as an operational factor affecting nominal embedment spacings rather than the design of penetrators.

It is noted that temperature effects would, in general, be less pronounced if alternative smaller diameter HLW canisters were adopted.

2.6.11 Thermal properties of penetrator materials

The peak temperatures reached by emplaced penetrators for a given heat output is governed by the thermal conductivity of the sediment. The thermal conductivity of penetrator construction materials is therefore not important.

2.6.12 Irradiation properties of penetrator materials

Only materials with good radiation resistance have been considered for penetrator construction.

A penetrator design giving better attenuation of HLW radiation than otherwise equivalent options because of the good shielding properties, or necessary thickness, or both, of its materials including fillers would be favoured. In many instances this would be advantageous with regard to the engineered barrier performance by reducing the ionisation of sediment pore water around embedded penetrators. The disposal of greater age HLW has a similar advantage because of radiation output decay.

The special case of thick walled self shielding penetrator designs is discussed in 2.7.7.

2.6.13 Corrosion product effects

Further research is needed to confirm that corrosion products would have no adverse effects on sediment radiological isolation properties or HLW leach rates.

2.6.14 Pre-disposal loads

Pre-disposal load is used to signify any load imposed on a penetrator from the time of component fabrication to the release from the disposal ship. Handling loads, transport vibrations and the effects of any mishaps such as lifting accidents are examples.

Proposed penetrators would have to be capable of resisting certain categories of accident loads. However, it has been assumed that it is not necessary to make special allowance for them at this early stage because penetrator options have been developed taking into account seabed embedment and hydrostatic loads and would thus be expected to be sufficiently robust.

Pre-disposal loads are also essentially controllable. For example, a lifting operation could be revised by the relocation or the use of more support points if it was considered necessary to reduce the stresses induced in a particular penetrator design. Similarly, assembly and transport procedures could be chosen to limit the height to which penetrators have to be lifted to reduce the possible magnitude of accidental loads for postulated mishandling incidents and other contingencies.

In this way, pre-disposal loads can be seen as influencing disposal operation planning rather than fundamental penetrator features though a selection of subsidiary design considerations is as follows.

a) generally it is unlikely that low temperature property requirements such as ductility would be needed for penetrator materials given the initial HLW heat output and the steady seabed ambient temperature of about 4°C. However, some degree of low temperature ductility should be specified for penetrator components so that they could be transported to the assembly plant during a spell of freezing weather without the danger of cracking .

b) tail fins would be more vulnerable to deformation than other parts of penetrators during a minor handling accident. Rather than designing massive fins to resist accidental loads it has been presumed that, if necessary, they would be protected by temporary guards from time of fabrication until the release of the penetrator from the disposal ship.

2.6.15 Disposal loads

Loads following the release of a penetrator from a ship are described as disposal loads. They would include hydrodynamic loads superimposed on increasing hydrostatic pressure during free-fall ocean descent followed by impact and penetration resistance loads during embedment into the seabed. Thereafter, embedded penetrators would be under essentially uniform hydrostatic pressure combined with surrounding sediment loads which may vary as a function of the HLW heat generation.

Hydrodynamic loads are relatively insignificant for the penetrator body but have been estimated for the purposes of fin design in Appendix B.

With respect to hydrostatic pressure, designs would be preferred with sufficiently thick walls to prevent buckling if there was a local area of unsupported penetrator wall.

2.6.16 Penetrator internal stresses

Internal material stresses such as residual stresses due to fabrication procedures, welding, temperature stresses and stresses due to flexure under external loads could be detrimental to penetrator performance. The stresses would be of most significance if it were possible for them to initiate local corrosion of the engineered barrier by the stress corrosion cracking mechanism discussed in 2.7.4.

The avoidance of sudden stress level changes in penetrator bodies is seen as an advisable precaution even if occurring outside the engineered barrier. Thus penetrator nose and tail pieces have been shown with smooth shape and thickness transformations.

2.6.17 Site usage efficiency

Site usage efficiency is discussed in 8.0 as an operational factor in terms of the nominal spacing between penetrators and takes account of thermal effects and emplacement accuracies.

The important principle for penetrator design is that the number of canisters carried should be the maximum consistent with other performance requirements and practical limitations. This reduces the number of penetrators needed for the disposal of a given volume of HLW so requiring a correspondingly smaller disposal site area given that nominal emplacement spacing is not much influenced by penetrator size.

It is difficult to strike an absolute balance between penetrator canister capacity maximisation for site usage efficiency and penetrator length and weight limitation for other requirements such as handling and transport. In this study, a one to twelve canister capacity option range was considered on the assumption that the optimum capacity for any particular type of penetrator would lie between the two extremes.

It is noted that, other things being equal, site usage efficiency would be reduced if alternative smaller diameter HLW canisters were adopted.

2.6.18 Penetrator instrumentation

A requirement to provide a compartment on penetrators for transponders or other equipment has been assumed. The main purpose of such equipment would be to help locate any penetrators lost at sea after a shipping accident. It would also enable the free-fall descent and embedment behaviour to be monitored for all or a sample number of penetrators during disposal. Some equipment for general environmental monitoring may also be required.

In order to avoid the problem of collapse under hydrostatic pressure it has been assumed that an open compartment would be provided together with instruments designed for water immersed operation.

2.6.19 Cost effectiveness

The objective of the study is to identify possible penetrator options which might be technically feasible as it is far too early to undertake cost analyses to optimise designs. Nevertheless, cost effectiveness, is one aspect that has been compared for alternative penetrator designs.

It is felt that, other things being equal, moderately increased disposal costs would result if alternative smaller diameter HLW canisters were adopted.

2.6.20 Penetrator design ramifications

The secondary performance requirements have limited influence on penetrator design compared with the three primary requirements for waste form compatibility, engineered barrier design life and embedment depth attainment. The desirability of improving site usage efficiency is an argument for large multi-canister penetrators but these were already favoured for enhancing embedment depth. However, the secondary requirements do include fabrication, handling, transport and thermal effect considerations which limit the maximum practical size for a penetrator.

2.7 Candidate Materials

Selections of material types and thicknesses and all the other decisions during the production of a specific penetrator design are vulnerable to question, review and amendment as the result of subsequent research and development programmes on, for example, corrosion behaviour. Indeed there is merit in not adopting rigid conceptions of penetrator design and construction at too early a stage so that flexibility is retained to respond to new research findings.

The following development of specific designs is, therefore, not held out to be the approach that would be adopted if an HLW penetrator disposal operation was commissioned because of the impracticality of making predictions for so far in the future. In addition, there is a great amount of further scientific research work that would have to be undertaken to give the necessary basis for detailed engineering designs.

The selection of appropriate penetrator materials is most influenced by the need to provide an engineered barrier. This restricts the material type and thickness combinations to those with adequate corrosion performance and behaviour under hydrostatic pressure for at least 500 years, as discussed in 2.4.4.

The embedment performance criterion has a lesser significance but favours the use of dense materials in relatively large thicknesses to enhance penetration depths into the seabed. The resistance of seabed impact loads is also discussed. The satisfaction of the secondary performance requirements listed in 2.6 is not detailed for every proposal because only options with a likelihood of meeting a full performance specification have been described in this report.

2.7.1 Overpack and penetrator body construction

The basic construction concepts of overpacks and penetrator bodies were introduced in 2.4.1 and 2.4.2 so that the discussion of performance requirements could be related to currently envisaged forms for penetrators. The four construction options are:-

a) corrosion resistant
b) corrosion allowance
c) pressure resistant
d) radiation shielding

These main options are shown in Figures 5,6,7 and 8 and will now be described in more detail.

2.7.2 General material and jointing considerations

Penetrator disposal would entail seabed impact loads that would not be present, or not present to the same extent, with other HLW disposal options. Therefore, only metals with good ductility to resist these loads have been considered for penetrator construction. The use of plastics is perhaps a possibility for future evaluation but their use for large scale industrial purposes analogous to proposed penetrators is so limited as to preclude a valid comparison with metal construction. Furthermore, plastics are not under serious investigation in current HLW disposal materials research programmes so further restricting the availability of information on them.

The question of sealing an overpack or penetrator body around a HLW canister also affects the choice of materials. Any connection between overpack or penetrator body components acting as the engineered barrier must not fail within the specified containment design life. Many jointing methods exist. Welding, screw coupling, heat shrinking or other friction type collars, gluing , clamping and bolting are some of the options.

Potential advantages and drawbacks can be envisaged for all jointing processes. For example, screw couplings could be easily assembled but the possibility of thread crevice corrosion products causing stresses which might accelerate other corrosion failure modes would have to be assessed.

Welding, despite the metallurgical changes it induces in components, is on current evidence felt to be the most likely approach to satisfy all performance requirements. It has, therefore, been assumed as the jointing process for penetrators. Thus only materials suitable for welding have been selected as candidates at this stage for penetrator components.

2.7.3 Material selection categories

Feasibility studies of HLW geological disposal schemes are assessing two main categories of metals as possible alternative approaches to satisfying the engineered barrier requirement (Ref. 25). They are:-

a) metals with very slow corrosion rates in the disposal environment. They are typically very expensive engineering materials such as titanium but it is possible that only relatively thin walled overpacks would be required so that their cost would not be prohibitive. This category of metals is known as corrosion resistant.

b) metals with significant but predictable corrosion rates such that it is feasible to provide sufficient thickness to remain unbreached for the desired containment period. This category is called corrosion allowance and essentially comprises weldable carbon steels. Cast irons could be considered if welding was eliminated or its difficulties overcome.

Many other metals, such as copper, have properties intermediate between corrosion resistant and corrosion allowance. These have not been considered as they represent neither of the performance bounds of interest.

Tables 2.1, 2.2 and 2.3 summarise the corrosion behaviour and certain properties of some materials which have featured in HLW research programmes. It should be noted that not all the materials, particularly some of the stainless steels and high nickel alloys, have been proposed for the engineered barrier function but have been listed because they are considered as possible HLW canister materials.

The hydrostatic pressures and embedment loads are a significant feature of the penetrator option and introduce structural performance requirements far in excess of those normally considered for alternative disposal in continental rock formations. These loads are a crucial factor both for material thickness selection and penetrator construction in that the more economic thinner walled design options developed in 2.8 would rely on the filling of all internal voids to prevent crushing of the penetrator body. Some materials which have been proposed for void filling are listed in Table 2.4.

Therefore, after the brief review of corrosion mechanisms in 2.7.4 , the necessary thicknesses or properties of materials for overpack and penetrator design options are discussed under four headings.

a) corrosion resistant metal thickness

b) corrosion allowance metal thickness

c) structural performance metal thickness

d) void filling material selection

2.7.4 Corrosion mechanisms

All metals other than gold and platinum will corrode in the seabed but, in doing so, some form a stable passive surface layer which inhibits subsequent degradation of the underlying metal. These are the corrosion resistant metals whereas the corrosion allowance metals such as steel have corrosion products which do not form a stable protective surface zone.

The seabed environment and the possible alteration of its characteristics over the containment design life are critical to corrosion predictions. For example, corrosion resistant engineered barriers might be expected to survive for as long as the disposal environment conditions remained benign, which could be far longer than the containment design life. Alternatively, some unknown long term corrosion mechanism such as protective film breakdown or an adverse transformation of the seabed environment such as could be caused by unpredicted biological activity could quickly destroy the barrier at some time before the end of its required design life.

The seabed disposal study areas have been selected partly because the clay-like nature of their sediments restrict pore water movements. Water carrying dissolved oxygen and oxygen which could be ionised by radiation from the HLW would be the principal factor determining the corrosion rate of a carbon steel penetrator. Thus the restricted pore water supply might be expected to drastically reduce the rates of steel corrosion. There is still a risk of adverse environment changes or unknown corrosion reactions accelerating the degradation of a steel penetrator. However, such possibilities should be easier to evaluate because the corrosion theories for steel are generally better developed and more certain than the understanding of protective film mechanisms for corrosion resistant metals.

Therefore, it should eventually be possible to predict the containment design life of steels with more confidence than for corrosion resistant engineered barriers. A disadvantage of steels is that they will probably have an ultimate life which is unlikely to be exceeded even if no adverse environment changes occur whereas the possibility of corrosion resistant options achieving significantly enhanced durations in benign conditions has already been mentioned.

In addition to uniform general attack, all metals are susceptible to some localised forms of corrosion which can include:-

- pitting attack
- crevice corrosion
- galvanic action (bimetallic)
- intergranular corrosion
- stress corrosion cracking (SCC)

General and most local corrosion mechanisms progress at rates for which appropriate metal thicknesses can be selected to avoid corrosion failure within a specified design lifetime.

Some mechanisms, principally SCC, once initiated, develop relatively quickly. It is economically impractical to provide large material thicknesses to allow for SCC so it is imperative to guard against its initiation in the first place. This requires the avoidance of adverse stresses by appropriate design and construction of components and the selection of materials which are least susceptible to SCC. It should be noted that the residual stresses at welds, particularly if relieving heat treatments are not undertaken, are potential areas for SCC initiation.

It would be inappropriate for this report to include a detailed review of the very large amount of research work which is being undertaken as a necessary adjunct to HLW disposal feasibility studies. Instead, further information may be found in publications of the SWG Waste Form and Canister Task Group (Ref. 12) and national research establishments or in various review papers (Refs. 16, 25 and 26).

2.7.5 Corrosion resistant metal thickness

Titanium alloys have so far been found to exhibit very favourable corrosion properties for HLW disposal engineered barriers. They have been identified as the most promising or been included in short lists of corrosion resistant candidate materials in the research programmes of several nations including European Community countries, Canada and the USA (Refs. 26, 27 and 28).

The behaviour of corrosion resistant materials in an environment which is changing, for example due to the effects of radiation and heat, is not well understood. Furthermore, there is insufficient data to allow corrosion rates to be predicted accurately over a 500 year period. Although this is also true for corrosion allowance materials, the difficulties in measuring very low corrosion rates exacerbate the problem in the case of corrosion resistant alloys.

However, a thickness of 5mm (Ref. 25) has become widely accepted on current evidence as offering the possibility of a design life in excess of 500 to 1000 years and has been assumed for the purposes of this study.

Corrosion resistant materials have been proposed for the engineered barrier function but rarely for structural functions such as embedment load resistance because of the very high cost of using them in the thicknesses, perhaps 25mm, that would be necessary. Therefore, in this study, they have only been considered for HLW canister overpacks. The overpacked canisters would be carried in a penetrator body of cheaper material for embedment in the seabed as discussed in 2.7.7. This approach has been called the corrosion resistant option and is shown in Figure 5.

The overpacks would have to resist hydrostatic pressure without collapsing during the containment design life so voids within the canister and overpack would have to be filled as discussed in 2.7.8.

2.7.6 Corrosion allowance metal thickness

Weldable steels are considered for the performance of a corrosion allowance engineered barrier function. However, the thickness of steel that would be needed for a 500 year design life is not at all clear from published results of corrosion studies.

Conservative assumptions regarding the corrosion environment have been made in research studies but the extrapolation of current experimental data to long term performance remains an uncertain exercise.

Nevertheless, a value of 75mm has been chosen for the purposes of this study. It has been called the corrosion allowance option and is shown in Figure 6.

The selection of a 75mm steel thickness is based on the following considerations:

a) general corrosion predictions which account for restricted water borne oxygen supply but which do not necessarily allow for temperature or some seabed effects range from 5 to 50mm. Corresponding local corrosion such as pitting depth predictions vary from severe (six times general corrosion) to moderate (similar or less than general corrosion) but researchers generally consider that the moderate depth estimates are the more realistic. The predicted depths of general and local corrosion have to be added to arrive at the necessary corrosion allowance thickness. The 75mm assumed for this study is thus of the same order as current, albeit preliminary, predictions.

b) a very approximate estimation of likely overpack costs presented in Table 2.5 indicates that a 75mm wall thickness design in steel could be marginally more expensive than a 5mm titanium alloy wall thickness overpack. The use of a thicker than 75mm steel engineered barrier would thus not be economically justified if a titanium-alloy overpack had a similar performance. This cost argument is further developed with particular reference to penetrators in 2.8.3.

A 75mm steel wall thickness engineered barrier could not be expected to resist hydrostatic pressure without plastic deformation during the containment design life unless all voids within it were filled as discussed in 2.7.8.

The possibility that internal voids could not be completely filled in practice has led to the concept of pressure resistant designs as discussed in 2.7.7.

Methods of improving the corrosion life of a steel containment to reduce its necessary thickness have been suggested by certain researchers. The effectiveness of galvanising or other surface protective coatings is being studied in Belgium whilst the possibility of small copper, chromium, manganese or molybdenum alloying additions to improve the corrosion performance of the the steel itself by a factor of perhaps two has been reported (Ref. 27). Such developments may affect criteria for proposed penetrators in the future. They have not been considered in detail in this study because they would not result in the adoption of a more conservative design approach.

2.7.7 Structural performance metal thicknesses

Two basic types of load are considered to govern penetrator structural design, namely:-

a) handling and seabed embedment loads (pre-disposal load combinations).

b) hydrostatic pressure (disposal load case).

Significant thicknesses of material are necessary to carry either load type so special corrosion resistant metals would be unsuitable candidates on the grounds of cost. Thus various thicknesses of steel are proposed in this study as the result of preliminary structural performance evaluations.

A 25mm penetrator wall thickness is considered sufficient to resist handling and seabed embedment loads. A penetrator body along these lines would be necessary for the disposal of HLW in corrosion resistant overpacks as illustrated in Figure 5.

Handling and seabed embedment load considerations do not affect the other options.

It is possible that internal penetrator, overpack and canister voids could not be completely filled in practice. If so, the corrosion resistant and 75mm steel thickness corrosion allowance designs would not be feasible. Thus a concept of a pressure resisting option has been evolved.

A 125mm carbon steel wall thickness is considered to be just sufficient to resist hydrostatic pressures without general yielding of the penetrator body. This assumes the use of steel with a yield strength of about 200 N/mm^2 which is typical for general engineering steels.

To the outside of this 125mm steel wall thickness must be added a corrosion allowance of either 5mm of corrosion resistant material such as a titanium alloy or 75mm of extra steel such that the penetrator could continue to resist the hydrostatic pressure without yielding during the 500 year containment design life specification.

The latter approach of 200mm thick monolithic steel is preferred for its relative simplicity although the non-uniformity of corrosion means that after 500 years most areas of the penetrator wall would be expected to be more than 125mm thick. It has been designated the pressure resistant option and is illustrated in Figure 7.

The bimetallic construction of a titanium clad 125mm thick steel wall body would be a practicable solution particularly if weight saving was necessary or research concluded that far more than 75mm of steel would be needed for a 500 year corrosion allowance. However, this design has not been evaluated in detail in this study because there are no definite operational or performance benefits that would justify the complication of its fabrication.

A third containment steel wall thickness of 300mm has been evaluated. Its performance would be essentially similar to that of the pressure resisting option discussed above but would have enhanced radiation shielding. It is possible that this would be advantageous for pre-disposal handling operations as the external radiation levels would satisfy the IAEA regulations for the transport of packages containing radioactive materials.

This design has been called the radiation shielding option and is shown in Figure 8. However, it has not been evaluated in great detail because a preliminary assessment indicated that any possible handling advantages would be outmatched by the drawbacks of increased cost, weight and fabrication complexity.

Internal void filling is not strictly necessary from the point of view of structural performance for the 200 and 300mm wall thickness pressure resistant and radiation shielding options. However, consideration of other factors such as performance enhancement beyond the 500 year containment design life leads to the conclusion that void filling should be strongly recommended and is discussed below.

2.7.8 Void filling material selection

In general it is considered that voids in pressure resisting penetrators would preferably be filled whilst all voids within engineered barriers of designs designated as non-pressure resisting would have to be filled.

Many of the factors involved with ocean disposal are similar to those for geological disposal of HLW on land. However, engineered barrier structural design is an important exception because external pressures at the envisaged ocean depths of up to 6km are much higher than would be experienced in nominally 1km deep generic repository designs which are currently postulated as options for land disposal.

Much thicker walled and thus more expensive penetrators would be necessary for pressure resisting designs with unfilled voids than for so called non-pressure resistant designs requiring void filling to carry the very high external loads involved with embedment in the seabed. The question of void filling is thus very significant for ocean disposal.

The most detailed assessment to date has probably been made by the Canadians who have selected lead, zinc, aluminium-silicon alloys and glass, sand, bauxite or steel shot pellets as an initial range of candidate filler materials (Refs. 29 and 30). The Canadians have also initiated experimental programmes to contrast pressure resisting and non-pressure resisting, that is void filled, overpack designs for disposal of HLW in land repositories (Ref. 31).

It is important to note that the practicality of fully effective filling of overpack or penetrator voids cannot be predicted with certainty at this stage. Detailed evaluations and trials of void filling techniques would, therefore, be necessary in order to assess the feasibility of non-pressure resisting penetrator options.

A representative list and brief evaluation of candidate filler materials is given in Table 2.4. The ullages in HLW canisters and clearances around canisters inside penetrator bodies are the two types of void filling that would be entailed. Filling with molten lead has been chosen as a reference assumption in advance of a comprehensive research programme to identify the most appropriate material.

An ideal filler would be initially fluid at room temperature for injection into voids before curing to a solid.

Alternatively, it has been assumed that a relatively low melting point material such as lead could be satisfactorily poured in its molten state on the basis that solidification shrinkage difficulties could be avoided by vibration or controlled cooling of penetrator bodies during void filling.

It is likely that canister ullage and penetrator body void filling would be undertaken separately. Also more than one stage could be most appropriate for penetrator body void filling. For example, Figure 37 illustrates a procedure whereby nose voids would be filled after penetrator body voids. If low melting point fillers were used in these circumstances, it is probable that alloys with successively lower melting points would be used at each stage. Therefore, canister ullage fillers would not be remelted by subsequent filling of penetrator body voids with lower melting point materials. The different melting points of the lead-magnesium alloy system is just one example of how this could be achieved.

Consideration of other aspects of penetrator assembly and performance specifications identifies further essential and desirable filler material characteristics in addition to structural properties including ductility as follows:-

a) compatibility with welding procedure and avoidance of weld contamination if welding adopted for completion of penetrators during assembly

b) inspectability so that any unsatisfactory void filling can be detected by non destructive examination during penetrator assembly

c) low volumetric change on solidification and during subsequent temperature variations

d) resistance to irradiation

e) compatibility with other penetrator materials

f) long term chemical and physical stability in excess of any containment design life specified for penetrators

g) low corrosion rate to prolong the engineered barrier containment of HLW

h) higher position in galvanic series than the metal used for the engineered barrier as discussed further in 2.8.3.

Finally it is emphasised that the materials and construction of the reference design of penetrator shown in Figure 15 were selected for the range of considerations introduced to the feasibility study and are not necessarily the optimum features which would be identified by a detailed planning assessment.

It should, therefore, be remembered that the assumption of lead as the filler material for the reference design is purely conjectural at this stage because of the paucity of research based recommendations. Consequently, two important aspects, namely density and cost, of the choice of lead as the reference filler material must be borne in mind.

The significance of density is clear from Figures 9 to 12 which contrast predicted seabed embedment depths for penetrators filled with lead and with much less dense cementitious grout. Significantly enhanced embedment performances are predicted for lead filled penetrators because they would be so much heavier. For example, a 5 canister capacity corrosion allowance penetrator filled with lead would be 40% heavier than a similarly proportioned penetrator with grout filled voids.

Filler materials could be expected to account for a considerable proportion of the total cost of manufactured penetrators particularly if a metallic filler such as lead were used. Table 2.4 gives a comparison of the approximate raw material costs per unit volume for the low melting point metals and alloys. It is more difficult to contrast the costs of metal fillers with non-metallic alternatives. However, in very general terms the raw material cost per unit volume of lead is currently about five times that of a typical cementitious grout and this does not allow for the significant energy cost that would be incurred by melting lead for injection into penetrator voids. It is, therefore, likely that non-metallic fillers such as grouts or plastics could at some future date be preferred on the grounds of economy if research shows that the use of, say, lead would have no particular technical benefits.

2.8 Candidate Penetrator Designs

The material types and thicknesses required for ocean disposal containment of HLW were reviewed in 2.7 in general terms without particular reference to penetrator designs. Only one aspect was discussed which would be needed for penetrator construction but not needed for a plain overpack intended to provide the engineered barrier if, say, it was lowered into a pre-drilled borehole in the seabed. The exception was the mention in 2.7.7 of the need for a 25mm wall thickness steel tube to resist handling and embedment loads for the corrosion resistant design option. Therefore, most of the preceding discussion has been equally appropriate to ocean disposal overpack design as to penetrator design. The incorporation in terms of specific penetrator construction proposals of the features necessary for HLW ocean disposal programmes is now reviewed.

2.8.1 Feasibility criteria

Questions of limitations of length, weight and component fabrication are introduced during the development of penetrator designs. Certain secondary performance requirements are affected and these are discussed when they concern a specific design option. However, reference should be made to 2.2.1 for the feasibility criteria which have been applied to test the general practicality of the proposed designs.

2.8.2 Overpack elimination

Corrosion resistant overpacks should be seen as a special case because it would almost certainly be uneconomic to make them with sufficient wall thickness to perform the structural function of a penetrator body to absorb embedment loads. Therefore, a composite penetrator design option has been proposed with canisters contained in individual corrosion resistant overpacks enclosed in a 25mm wall thickness steel penetrator body.

Using corrosion allowance materials such as steel throughout does, on the other hand, open up alternative penetrator construction approaches. The engineered barrier function could either be incorporated as monolithic with the penetrator body or consist of individual canister overpacks that would be subsequently loaded or joined into a penetrator.

The multi-canister capacity monolithic body approach appears to offer significant advantages in terms of component fabrication and penetrator assembly. Although remotely controlled loading of canisters and void filling in a specially built radiation shielded assembly plant could prove more involved for multi-canister penetrators, it would only require one weld to complete a penetrator. Alternatively, single canister overpacks would have to be individually welded to seal the engineered barrier and then have further joins made to build them into penetrators. Adjacent overpack ends would then form bulkheads in an assembled penetrator. These would add weight and length with consequent further cost and handling implications.

As a result of the assessments above, designs incorporating separate corrosion allowance overpacks after the manner shown in Figure 6 have been eliminated from further consideration in favour of the monolithic types of construction shown in Figures 7 and 8.

2.8.3 Penetrator design options

Alternative penetrator cross-sections are considered here. Lengths and corresponding canister capacities for penetrators are dealt with in 2.8.4.

The main options selected for evaluation are shown in Figure 13. The main difference between the alternatives is the variation in wall thickness of the steel body as follows:-

a) the minimum wall thickness tube design of 25mm corresponds to a corrosion resistant engineered barrier design.

b) the reference wall thickness tube design of 75mm corresponds to a corrosion allowance engineered barrier design.

c) the maximum wall thickness tube design of 200mm corresponds to a pressure resistant engineered barrier design.

The penetrator tubes have been chosen with a nominal internal diameter of 500mm. This would leave a 35mm wide annular gap around the HLW canisters for the injection of void filling material. In the case of the 25mm wall thickness tube, the annular gap would also have to accommodate thin corrosion resistant overpacks.

The enhanced embedment performance of the thicker walled and thus heavier options is reviewed in 2.8.6.

A point to note from the general comparison of options presented in Table 2.6 is the apparently similar material and fabrication cost of the 25mm wall corrosion resistant and 75mm wall corrosion allowance designs. This has been taken as one justification of the selection of a 75mm steel thickness for the 500 year containment design life corrosion allowance in advance of specific corroborative experimental data.

A desirable design criterion is to maintain an advantageous progression of increasingly cathodic materials from the outside to the inside of a penetrator. Thus any bimetallic corrosion will prolong the life of an inner metal layer following the local failure of an outer component. The achievement of this by the proposed penetrator designs can be seen from the galvanic series of materials in seawater given in Table 2.7.

It is stressed that the penetrator wall thicknesses chosen for evaluation are nominal values based only on preliminary experimental data and research results. More appropriate wall thicknesses will no doubt be identified as more information becomes available but it is expected that they will still lie between the extremes selected in this study.

2.8.4 Penetrator length options

The extreme penetrator lengths chosen for consideration in this study are:-

a) minimum length with capacity for one canister

b) maximum length with capacity for twelve canisters

As discussed in 2.6.20, large multi-canister penetrators would be favoured for good site usage efficiency and enhanced embedment depths. However, the maximum practical length for a penetrator is set by consideration of factors such as fabrication and transport.

The most appropriate penetrator length is felt to be between the two extremes given above. Figure 13 therefore illustrates three options:-

a) minimum length for one canister

b) reference length for five canisters

c) maximum length for twelve canisters.

2.8.5 Nose, tail and fin design

Some secondary factors such as hydrodynamic design and wall thickness transformations were discussed in 2.6. Otherwise very few fundamental constraints have been identified for nose, tail or fin design apart from practicality of fabrication.

A nominal height to diameter ratio of 1.5:1 for nose and tail pieces has been adopted (Ref. 7) but this should not be regarded as a definite requirement.

Fin sizing has been guided by output from the hydrodynamic computer model described in Appendix B.

2.8.6 Penetrator performance comparisons

All the proposed penetrator designs have been selected for their ability to satisfy the performance requirements discussed in 2.3 to 2.6. In particular, all the options would be expected to achieve a 10m minimum seabed embedment and provide a 500 year minimum containment.

In theory then, there is little to chose between the options on the ground of minimum acceptable performance criteria. However, some options offer enhancements which have been summarised below.

Corrosion resistant engineered barrier designs could achieve extremely long containments beyond the specified minimum design life if disposal environment conditions remained benign.

Pressure resistant designs would be expected to have an enhanced containment life compared with corrosion allowance designs because the structural thickness of the penetrator body would probably afford some additional corrosion life even if hydrostatic pressure began to cause some deformation. However, the estimation of the likely containment life extension would be difficult and uncertain. The thick walled pressure resistant design also has a significant radiation attenuation effect and good heat dissipation performance which would be beneficial for reducing seabed environment changes.

The main performance parameter for discriminating between options is embedment depth enhancement. In general, the heavier a penetrator design, the greater the embedment that it could achieve.

2.8.7 Component transport

The need to transport penetrator components was discussed in 2.6.2. With regard to present day practices it should be noted that there would be complications for the road transport of penetrator bodies in the following cases:-

a) penetrators with capacities for more than about eight HLW canisters would exceed normal transport length restrictions.

b) 200mm wall thickness penetrators with capacities for more than about five HLW canisters would exceed normal transport weight restrictions.

For transportation by rail, no component size or weight restrictions would be expected for any of the penetrator options.

2.8.8 Penetrator transport

The conveyance of assembled penetrators in transport flasks was discussed in 2.6.5. With regard to present day practices it should be noted that handling capacity limitations would be exceeded for penetrator and IAEA transport flask combinations in the following cases:-

a) 25 and 75mm wall thickness penetrators with capacities for more than about five HLW canisters.

b) 200mm wall thickness penetrators with capacities for more than about three HLW canisters.

2.9 Conclusions

The following remarks summarise the main topics which have been assessed and penetrator construction options which have been identified in this section.

Three primary performance requirements have been identified for penetrators, being:-

a) compatibility with the form adopted for vitrified HLW such as canister shape. See 2.3.

b) fulfilment of an engineered barrier function to isolate the HLW from the seabed sediments for a minimum specified containment period nominally set at 500 years. See 2.4.

c) achievement of a minimum specified tail embedment depth into the seabed nominally set at 10m. See 2.5.

Various secondary factors have also been discussed, namely:-

d) component fabrication, see 2.6.1

e) component transport, see 2.6.2

f) penetrator assembly, see 2.6.3

g) void filling, see 2.6.4

h) penetrator transport, see 2.6.5

i) penetrator storage, see 2.6.6

j) penetrator handling, see 2.6.7

k) pre-disposal contingencies, see 2.6.8

l) hydrodynamic speed and trajectory, see 2.6.9

m) peak sediment temperature rise, see 2.6.10

n) thermal properties of penetrator materials, see 2.6.11

o) irradiation properties of penetrator materials, see 2.6.12

p) corrosion product effects, see 2.6.13

q) pre-disposal loads, see 2.6.14

r) disposal loads, see 2.6.15

s) penetrator internal stresses, see 2.6.16

t) site usage efficiency, see 2.6.17

u) penetrator instrumentation, see 2.6.18

v) cost effectiveness, see 2.6.19

No fundamental insurmountable problems associated with satisfying any of these requirements have been identified in the course of the study.

The fulfilment of the engineered barrier function has been found to have most significance because of the constraints imposed on the selection of appropriate material types and thicknesses for penetrators. The resistance of the very high external pressures inherent with ocean disposal and the possible advantages for pre-disposal handling of thick walled penetrators to reduce radiation below permitted maximum levels for exposure of workers have also been discussed. As a result, four basic design options as shown in Figures 5 to 8 have been developed, namely:-

a) corrosion resistant

b) corrosion allowance

c) pressure resistant

d) radiation shielding

It is considered that corrosion behaviour research has not yet reached the stage when definite decisions can be made regarding the optimum material types and thicknesses for penetrators. However, it is concluded that current trends in the results of laboratory trials increasingly indicate that reasonable thicknesses of corrosion allowance materials such as steel could prove to be acceptable for HLW containment. On the other hand, it has been shown by the very approximate analysis presented in Table 2.6 that a corrosion resistant design incorporating a 5mm thick Titanium engineered barrier could be more cost effective than corrosion allowance designs entailing steel wall thicknesses in excess of 75mm. Therefore, the identification of a steel thickness to give a specified containment design life in the conditions of the deep ocean seabed should remain an important objective of corrosion research programmes. Certain tentative assumptions about the material thicknesses appropriate for the various alternative design options have been made for the purpose of this study and are given in Table 2.5.

3.0 DISPOSAL OPERATIONS

3.1 Introduction

This chapter of the study considers the operations involved in taking the waste from the interim store to the point at which the penetrator is released to fall to the seabed.

Several alternative sequences of operations are proposed and their feasibilities are assessed. Those which are worthy of more detailed attention are identified for further consideration in Part 2.

3.2 Principal Assumptions

The study is based upon the disposal of 85m³ of HLW annually; but see Appendix A and 3.13 for the effects of increased volumes.

It is assumed that the canisters of vitrified HLW are produced at an inland site. This ensures that the problems of transportation to the dock are not overlooked.

The canisters will be collected from a dry vitrified product store when the waste is at least 50 years old (ex reactor) unless otherwise stated.

The waste is transported to the disposal site by a surface ship.

It is assumed that the canisters will not have deteriorated during the period of interim storage and can be retrieved and handled by means similar to those used when the waste was vitrified and put into the store.

3.3 General Discussion

All the sequences of operations proposed in this study comprise of land transport, sea transport, penetrator component manufacture and assembly, and penetrator launch phases.

Six generic options have been considered and the salient features of these are shown in Figure 14. The six all assume that HLW is produced at the same inland site and is all eventually disposed of at a common deep ocean location. The main variables are the nature of the coastal facility and the type of disposal ship.

The system consists of a number of stages involving transport, handling, manufacture and storage. These stages may not all be required in any one system and may not appear in the order set out as follows:-

a) canister storage
b) land transport
c) overpacking

d) penetrator assembly
e) dockside storage
f) transfer to ship
g) sea transport
h) penetrator launch

Overpack and penetrator component manufacture are not included in the above list as the operations involved would not entail the presence of HLW.

3.4 Transport

3.4.1 Current and proposed regulations

For the purposes of this study it has been presumed that the procedures for any transport of HLW would be in accordance with the IAEA Transport Regulations (Ref. 32). Procedures for the movement and storage of HLW within a nuclear establishment have to be agreed with the appropriate licensing authority and, therefore, are not within the scope of the IAEA Transport Regulations.

The existing regulations would require HLW to be transported as a Type B package because of its high activity. They do not yet explicity cover the marine transport of penetrators.

3.4.2 Current and proposed practices

It is felt that land carriage by rail will be identified as the most practical, safe and acceptable means of HLW transport but the possibility has not been ruled out that shipment by inland waterway could be an appropriate alternative.

Transport on public roads is less likely except in extreme circumstances involving, for example, rail access difficulties, because it is less safe and more inconvenient than by rail (Ref. 33). Identifying suitable road routes for the heavy loads proposed could also pose problems.

Carriage of HLW by rail, road or inland waterway would require the use of radiation shielding, heat dissipating, accident resistant transport flasks comparable to, but in many ways simpler than, those currently used for spent reactor fuel shipments. Thus, only relatively straightforward adaption of existing technology and transport procedures would be needed to meet the requirements of HLW transport.

An important aspect of current design practice is that spent fuel transport flasks more than satisfy the performance specifications of the IAEA Transport Regulations with regard to radiation shielding. This is to satisfy stringent exposure limits at handling facilities within nuclear establishments although flask design factors of safety also play a part. As a result radiation levels measured at the surfaces of flasks in current use are typically one or two orders of magnitude less than the maximum limit specified in the IAEA Transport Regulations.

It is argued that HLW transport flasks should have equally good shielding. This presumption particularly influences the consideration of what others (Ref. 34 and 35) have called "shielded" HLW containers. These are HLW canisters with nominal 300mm thick mild steel overpacks which reduce external radiation levels when the HLW is at least ten years old (ex reactor) to about the maximum limits permitted by the IAEA Transport Regulations. Nevertheless it has been assumed in this study that transport flasks would be used even for "shielded" HLW containers to ensure the much lower radiation exposures mentioned above.

The surface radiation levels for various thicknesses of steel shielding overpacks placed around HLW canisters are shown on Figure 4.

Spent fuel flasks of approximately 115t gross weight are already in service and it is considered likely that flasks of a similar weight would be a feasible means of transporting HLW.

3.5 Handling

3.5.1 Weight

The weight of the canister including HLW but no filling is approximately 480 kg.

The weight of an overpacked canister depends on the size and density of the overpack and the type of filler, if any, used.

The maximum gross weight of a fully assembled penetrator has been set at 100 tonnes. This limit has been established by the known availability of lifting gear and is approximately the weight of the larger current transport flasks.

3.5.2 Length

The principal restrictions are:

a) the interdependence of length and weight.

b) the ability to manoeuvre long lengths, which depends on the disposal operation chosen.

c) the structural handling restrictions. These may be overcome by increasing the load capacity of connections between overpacks or increasing penetrator shell thickness.

Although no absolute upper length limit for the penetrator has been established at this stage, a consideration of the above factors has led to the adoption of a 20m upper length limit.

3.5.3 Shielding

The need to protect the public and operatives from the effects of radiation is self evident. The required shielding can be achieved in the following ways:

- by direct shielding, permanently attached to the canister.
- by removable temporary shielding.
- by handling the canisters inside shielded cells using remote handling equipment.

Normally all three types of shielding are present to some extent or another and they would all be included in the radiation assessment of the process.

Permissible dose rates to both operators and the public are governed by various statutory authorities. However, the dose rates used for the design of new nuclear facilities are usually lower than the permissible rates by an order of magnitude. Typical current rates are 2 mSv/man/year from normal plant operations and 5 mSv/man/year total when exposures during plant maintenance are included. None of the suggested overpacks or penetrator bodies would be expected to provide sufficient shielding to give acceptable surface radiation levels for prolonged exposures of workers.

The conclusion reached is that additional shielding will be required, but that the beneficial effects of the overpack will reduce the extent of the additional shielding.

3.5.4 Temperature

The temperature before disposal can be calculated knowing the thickness of the overpack, the age of the waste and the ambient conditions.

IAEA Transport Regulations state that the maximum temperature of readily accessible surfaces during transport as a full load must not exceed 82°C in the shade under normal transport conditions. When not transported as a full load, the limit is 50°C.

3.6 Manufacturing Location

The fabrication of the overpack or penetrator body is most likely to be performed in a conventional workshop and the fabricated components transported to the penetrator assembly station. No radiation shielding would be required during the manufacture of components.

3.7 Assembly Location

Figure 14 sets out the possible locations where the component assembly may take place. It can be seen that both overpack and penetrator assembly could take place on board ship and the disadvantages of this are discussed later in this section.

Equally penetrator assembly could be situated at the canister source or at the dockside. It will be shown in Part 2 that the size of the penetrator affects this decision.

3.8 Storage

3.8.1 Interim storage

British Nuclear Fuels plc (BNFL) is building a dry surface store for filled HLW canisters at Sellafield. The canisters will be stored 10 high in vertical compartments. The natural convective cooling air flow will not come into contact with the canisters so there should be virtually no radioactive discharge, no contamination and no secondary waste.

3.8.2 Temporary holding storage

Overpacked canisters and assembled penetrators would also probably be kept in dry storage because of the simplicity of operation, lower costs and virtual elimination of the need to decontaminate.

3.9 Dockside Establishment

The minimum establishment would be a berth with a draught of 5m or greater, and a holding area for canisters or penetrators prior to loading. The present UK legislation, The Nuclear Installations Act 1965, does not require "storage incidental to transport" to be licensed. A dock and holding area for a few canisters or penetrators awaiting loading in transport flasks would therefore not be expected to require to be licensed as a nuclear site.

At the opposite extreme, the entire penetrator assembly could be located at the dockside. It would receive HLW canisters direct from the source and produce penetrators ready for loading and subsequent release. Such an establishment would certainly require a nuclear site licence.

A scheme for this option is contained in Part 2.

3.10 Shore to Ship Transfer

It is desirable that the storage conditions at the dockside and aboard ship are compatible. Handling should be minimised as this represents a potentially hazardous part of the operation.

Four transfer mechanisms will be discussed in Part 2 when the various ship design and transfer options are presented. At this stage only a very brief indication of four feasible alternatives is given.

3.10.1 Roll on roll off

Canisters or penetrators in land transport flasks would be conveyed on board via a loading ramp. Many suitable ramps have been developed and installed for comparable loading operations.

3.10.2 Crane

Canisters or penetrators in flasks could be lifted on board by dockside cranes. It is unusual to have cranes of greater than 40 tonne capacity so this would limit the size of penetrator (without purchasing a large crane). At Barrow, BNFL have recently commissioned a 150 tonne crane to lift fuel flasks.

3.10.3 Shielded tunnel

Unshielded canisters or penetrators could be moved inside a shielded tunnel built into the dockside at grade, or slightly below, to mate up with a door in the side of the ship set between the weather (top) deck and the water level. The mechanical handling equipment within the ship then positions the cargo, in special holds.

3.10.4 Underwater transfer

Capitalising on the good shielding and cooling properties of water, assembled penetrators could be transferred below the sea surface through a door in the ships hull to a flooded hold. There are several disadvantages with this method and these will be discussed in Part 2.

3.11 Ship Design and Operation

The only analagous existing marine transportation is the shipment of irradiated nuclear fuel. This is an entirely passive operation as the fuel remains sealed in transport flasks during the voyage. The many aspects of ship design will be discussed later but even the passive transportation of HLW requires many safety features to be incorporated into the ship design. The options which involve penetrator assembly on board ship lead to an unacceptable degree of ship sophistication, expense and hazard potential.

For the time being it is sufficient to mention a few salient features of ship design.

3.11.1 Design regulations

The early intercontinental transport of irradiated fuel was conducted using existing ships which had been adapted to carry fuel flasks. As the traffic increased new purpose designed ships were introduced. These incorporated many safety features to conform with the International Maritime Organisation dangerous goods code and the Japanese Ministry of Transport criteria. No particular UK legislation exists for ships carrying radioactive material, neither is there an upper limit on the total activity aboard.

3.11.2 Ship design features

The above regulations combined with good practice lead to several unusual design features. These include collision resistant side tanks around the hold, double bottoms to reduce grounding damage, twin engines and propellers, bow and stern thrusters for increased manoeuvrability and duplication of other important equipment.

3.11.3 Use of existing shipping

In the light of the above outline of the special requirements, it is unlikely that the conversion of existing ships is going to be either satisfactory or cost effective.

3.11.4 Minimum ship size

The ship must be sized to behave satisfactorily in extreme sea conditions. Based on the designs of spent fuel transport ships and conceptual studies of waste disposal ships a minimum length of 100m can be derived. Using conventional ship design ratios this leads to a 3000 DWT vessel.

To obtain the stability needed to assemble penetrators aboard would require a much larger ship whose corresponding increased draught would limit the range of ports it could enter.

3.11.5 Factory/disposal ships

A factory/disposal ship which included shielded facilities for canister overpacking and penetrator assembly would have to be very sophisticated, costly and likely to be under utilised. Specialist personnel to operate the factory and monitor the operation would have to be included amongst the crew. There would be particularly severe difficulties regarding the development of safe and reliable remote controlled equipment.

3.12 Penetrator Release

The interdependence between the loading, ship storage and release operations is very important. It would be beneficial if the mechanical handling equipment needed to receive and store the penetrator could be used again to release it.

Prior to release the ship would sail to the disposal site, positioning to an accuracy of about 30 metres using satellite navigation. By a combination of forward propulsion and the bow and stern thrusters, the ship would be able to keep station and the penetrator would then be positioned, inspected and released. As release rate is not expected to be a major consideration, the ship would be able to check that emplacement had taken place as planned before moving into position for the next release.

Four feasible generic release modes have been identified and three are developed in Part 2.

3.12.1 Over the side

This involves the use of a ship mounted slewing crane which would pick up the penetrators, place them in the water and release. The main draw-back is the lack of control during the manoeuvre and the difficulty of shielding the penetrator during the operation. This option is not developed in Part 2.

3.12.2 Launching ramp

The penetrator would be placed on an inclined ramp and released to slide down the ramp into the sea.

3.12.3 Through the hull

Penetrators stored or turned into a vertical position can be released through locks in the ships hull. This is preferred to the more complex operation of launching horizontally through locks or bottom opening doors.

3.12.4 Moonpools

Moonpools are becoming increasingly popular in ships used for underwater operations. The area of calm water within the ship's structure in direct contact with the sea would facilitate the disposal operation. To limit the motion of the moonpool it should preferably be located near the centre of the ship.

3.13 Disposal Operation Options

The numerous alternative sequences of operations for disposing of the HLW have been grouped into six possible options A,B,C,D,E,F as shown in Figure 14.

Because of the uncertainties in the European nuclear power programme it would be unwise to suggest a disposal operation method which is only practical if the maximum predicted waste volume of $400m^3$ annually is involved. It has been decided to design the method to handle the lower volume of $85m^3$ annually and suggest that the number of ships, flasks etc. would be increased to deal with larger volumes. This approach also permits the process to be established in more than one country.

The options will be discussed in the following sections in the light of the factors discussed in the preceding sections. Intermediate holding storage is not discussed with each option but it will be required between stages to provide a buffer.

3.13.1 Option A

The penetrators are assembled at the canister source and are taken to the dock in transport flasks which are brought directly aboard. Weight restrictions limit the penetrator size to 5 canisters. The main advantage is that a new nuclear site would not have to be found but it does place a restriction on the size of the penetrator. It also means that the flasks would be taken to sea and they may have to function both as transport and as release flasks, adding to their cost and complexity.

This option is developed further in Part 2.

3.13.2 Option B

Overpacking is performed at source and the penetrators are assembled at the dockside factory. Transportation to the dock does not limit the penetrator size.

It is concluded in 4.2 that individual overpacks are not advantageous so this option is not discussed further.

3.13.3 Option C

The canisters are overpacked at source but the penetrators are assembled aboard ship, severely complicating the ship design, exposing the difficult assembly operation to the rigours of the sea and splitting the operation between two sites.

This option is not favoured and is discounted.

3.13.4 Option D

The canisters are taken to the dock in a land transport flask. On this new licensed site the canisters are loaded into penetrators which are taken aboard ship in their finished form.

This option has the greatest number of technical advantages as all activities are concentrated at one land site. The major disadvantage compared with option A is that a large suitable site needs to be found and licensed.

This option is developed further in Part 2.

3.13.5 Option E

The canisters would be overpacked at the dock but the penetrators assembled at sea. Again assembly at sea is not favoured and this option is discounted.

3.13.6 Option F

In option F the canisters are transported directly to the ship which both overpacks and assembles penetrators. This results in the most complicated ship of all and is least favoured.

3.14 Conclusions

The operational aspects of taking the HLW from the initial source to releasing the penetrators at sea have been investigated. Six options (see Figure 14) have been presented demonstrating alternative ways of performing this task. There are many variations possible, but it is thought that these six options present a wide perspective and assist in establishing preferred schemes.

Only relatively straightforward extensions of existing technology are required to establish land based canister and penetrator assembly facilities. It is felt that the development costs, the equipment and ship capital charges and the general unsuitability of including the assembly facilities mean that factory ships cannot be justified.

Therefore, Disposal Options C, E and F, are not considered worthy of further investigation.

From these conceptual studies, it is concluded that options A, B and D are feasible. However, inevitable duplication of staff and facilities caused by splitting overpack and penetrator assembly in Option B detracts from this option and it is not considered further.

The choice between options A and D depends largely on the length of the penetrator. Shorter penetrators are easier to handle, transport and release. Longer penetrators have the probable advantage of being more cost effective. They require fewer nose cones and fins, fewer penetrator releases for a given number of canisters, less disposal area and they penetrate further into the seabed. However, it is not considered to be feasible to transport 20m long, assembled penetrators on the public road and rail network.

Finally, it is concluded that the operational efficiency is not sensitive to waste volumes greater than the $85m^3$ per annum assumed disposal rate but that volumes much less than this will cause under utilisation of equipment.

In Part 2, options A and D are discussed in greater detail and scheme designs are presented for the ships and docks.

4.0 SELECTED OPTIONS

4.1 Introduction

Part 1 of this study report has confirmed the preliminary engineering feasibility of the penetrator option for the disposal of HLW. The many factors which affect disposal have been investigated and discussed and, to date, no overriding reasons to doubt this conditional feasibility have been ascertained.

The following sections summarise the initial consideration of penetrator design options and disposal operation options, and then outlines the reference penetrator design option to be used in Part 2 as a basis for more detailed studies.

4.2 Summary of Penetrator Design Options

The study has reviewed possible alternative penetrator designs and tested them against the three principal and numerous secondary requirements which have been established in Section 2.

Of the three principal requirements the need to provide an engineered barrier around the emplaced HLW canisters has been found to be the most onerous stipulation. The other main requirements, namely, for the penetrator to be compatible with the waste form and to achieve a minimum tail embedment of 10m into the sediments are both readily achievable.

Manufacturing, transport, assembly and cost considerations determined that the barrier was best provided by a single penetrator body shell containing several canisters rather than by individually protecting the canisters.

The barrier is intended to prevent the ingress of sea water into the penetrator for the first 500 years after embedment. During this time, the temperature of the waste will have decayed to nearly ambient levels.

Four design variations have been developed covering a wide range of alternatives, all four satisfying the design conditions. As it is not practical to develop these designs fully, it is necessary to select one or two designs for further study.

Penetrator bodies made from thin skins of highly corrosion resistant materials, such as titanium alloys, require structural filling of the voids within the penetrator as the thin skin cannot resist the hydrostatic pressures alone. In addition, the uncertain corrosion rates, the difficulties of manufacture and the sensitivity of the material to mechanical damage, militate against this design and no further study has been undertaken.

In contrast, mild steel is a readily available material with adequate structural strength and good fabrication properties. Although the anticipated corrosion rates are much higher than for titanium, the rate is sufficiently low not to have to start with an unreasonable thickness.

Providing 300mm of mild steel is likely to give a good margin above the thicknesses needed for pressure and corrosion and goes much of the way to provide adequate radiation shielding. However, the consequent weight and cost have been shown not to be worthwhile and this option has not be considered further.

A steel shell approximately 75mm thick has been shown to be a potentially feasible way of providing 500 years containment when integrated into a penetrator body design, although further corrosion research would be required to confirm the metal thicknesses proposed. This option is considered in more detail in Part 2.

4.3 Summary of Disposal Operation Options

The various ways in which the vitrified waste could be successfully emplaced into the sediments using free fall penetrators have been studied and several different feasible operational methods have been identified.

Penetrator assembly has been found to be the operation which has the greatest effect on the operating system. Handling and transport constraints have been found to limit the size of the assembled penetrators which can be moved on the road and rail network. However, penetrators containing up to 5 canisters could be assembled at the vitrification works site and could be transported to the dock in flasks.

The assembly of penetrators onboard ship has been investigated but found to be of little merit as it is an inefficient use of equipment and manpower, and involves unnecessary risks.

4.4 Selected Reference Option

To provide a working hypothesis for Part 2 studies, a reference penetrator design option has been selected which incorporates all of the main desirable features highlighted by the study so far. It should be remembered that this is not an optimised design and should not be regarded as such.

The reference design would contain 5 HLW canisters in a single mild steel body shell of 75mm thickness. The penetrator shell could be manufactured in conventional workshops and would then be transported to the assembly place.

The assembled penetrator voids would be filled using a suitable filler which would act compositely with the waste and the penetrator body to resist the external forces. The penetrators would be transported to the release site by purpose designed ships fitted out to store and release them but not to provide assembly facilities.

The reference penetrator design is shown on Figure 15.

PART 2 - FURTHER STUDY OF SELECTED OPTIONS

5.0 PENETRATOR SPECIFICATION

5.1 Introduction

Part 1 reported the initial stages of the study which reviewed penetrator design options and disposal operation techniques and concluded by recommending a reference design of a 75mm thick mild steel penetrator shell containing 5 canisters as shown in Figure 15.

Part 2 takes the penetrator design and operations studies to a more detailed stage of development and considers also the problems which would be involved in manufacturing penetrators.

Site usage and management topics are discussed, and an estimate of disposal cost is made before a final assessment of feasibility.

Section 5 develops an outline specification for the manufacture of the reference penetrator.

5.2 Reference Penetrator Specification

5.2.1 Guide steel chemical composition

Corrosion research programmes have not progressed to a stage which would allow a steel chemical composition to be specified on the basis of optimised engineered barrier performance. Currently no definite data exists to show that one grade of steel has a particularly better seabed corrosion life than another. The possibility that small copper, chromium, manganese or molybdenum alloying additions would reduce steel corrosion rates has been suggested but remains to be confirmed by suitably detailed trials (Ref. 27). In any event, dramatic changes in manufacturing considerations would not be expected even if a steel was modified in such a relatively minor fashion.

At this stage, therefore, it is justifiable to evaluate penetrator construction in a steel specified to current engineering standards. Thus a weldable high purity steel with routine mechanical properties, appropriate to general engineering applications is proposed. A requirement for high purity stems essentially from the advantage of avoiding certain types of weld defect.

The selected material is described as a fully killed, weldable, high purity (low sulphur and phosphorus) electric quality steel produced by either a double slag or a single slag and secondary steel making process. Vacuum degassing would be a further requirement except, possibly, if the steel was to be sand or centrifugally cast for the reasons discussed in Appendix E.

A guide chemical composition for the proposed steel is presented and discussed in Appendix C.

5.2.2 Guide mechanical properties

As explained in 2.7.8, non pressure resisting options, including the reference design, rely on void filling to achieve composite structural action of the steel penetrator bodies with their contents. The absence of voids in the penetrator will minimise the possibility of structural failure due to hydrostatic pressures.

As no particular requirements for exceptional strength and ductility can be identified at this stage, typical values for general engineering steels have been assumed.

A guide mechanical properties specification has been taken as follows:-

410 N/mm² minimum tensile strength
200 N/mm² minimum yield strength
20% minimum elongation
25J minimum Charpy V-notch impact value at room temperature

When considered in conjunction with the guide chemical composition, these mechanical properties are similar to those for steels specified to British Standards (BS) as follows:-

a) BS 1501 grade 224 - 430RT for plates up to 150mm thick
b) BS 1503 grade 224 - 410 for forgings
c) BS 3100 grade A1 for castings
d) BS 4360 grade 43E for plates over 150mm thick

5.2.3 Heat treatments

Heat treatments are required to achieve the specified mechanical properties in a steel. A typical final heat treatment procedure for the guide chemical composition and mechanical properties of the steel would be normalizing followed by tempering.

Normalizing would involve heating to about 950°C followed by relatively slow cooling to give a fine grain structure.

Tempering would involve heating to about 650°C for around 30 minutes to obtain ductility and toughness. It could also be regarded as a stress relieving treatment for the steel being considered.

These temperatures could not be permitted once the HLW had been loaded into the penetrator on account of possible damage to the waste and any fillers.

5.2.4 Grain size

Small grain sizes in a steel are generally preferred because of the beneficial effect on properties such as toughness. A requirement for grain size to be finer than No. 7 measured to British Standard 4490 has thus been assumed as there are currently no indications that grain size would have a significant effect on seabed corrosion of a penetrator.

5.2.5 Tolerances

Aspects such as hydrodynamic behaviour and load resistance are insufficiently developed for penetrator manufacturing tolerance requirements to be fully evaluated. During this feasibility stage, therefore, requirements based on British Standard 5500 recommendations for construction of vessels to resist external pressure have been adopted as follows:-

External circumference : Co (nom) ± 5mm range not to be exceeded at any penetrator cross-section

Circularity = $\frac{Di(max)-Di(min)}{Do(nom)}$: not to exceed 0.01 at any penetrator cross-section

Additional nominal tolerance requirements that could be reasonably expected for the manufacture of penetrators have been identified as follows:-

Wall thicknesses: e ± 3mm range not to be exceeded

Lengths : L ± 10mm range not to be exceeded

Straightness : The maximum deviation from a straight line not to exceed 10mm over the full length nor 5mm over any 1000mm length of a penetrator tube

Where:

	Co (nom)	=	nominal outside circumference
	Do (nom)	=	nominal outside diameter
	Di (max)	=	max. actual internal diameter measured at a cross section
	Di (min)	=	min. actual internal diameter measured at same cross section as Di (max.)
	e	=	nominal wall thickness
	L	=	nominal length

When considered together, the application of these tolerances would, for example, permit about a ±8mm variation of the nominal 500mm internal diameters of penetrators.

5.2.6 Surface texture

There is no clear need for a particular surface texture to be specified to satisfy any aspect of penetrator performance. Although a moderately rough finish on the inside surfaces of penetrator bodies will improve bonding with the void filling material it is likely that sufficient adhesion will be achieved on relatively smooth surfaces anyway for adequate composite structural action.

The necessity to specify a surface texture stems instead from the need to facilitate non destructive examination (NDE) of penetrator components as several important NDE techniques require smooth surfaces. This is particularly true for surface defect detection methods but also simplifies certain aspects of ultrasonic testing for internal material defects.

A nominal surface texture has, therefore, been set at 12.5 micrometres Ra (N10) for 0.8mm meter cut off as assessed to British Standard 1134 Part 1. It is possible to achieve this standard on as-cast surfaces of moderately sized steel castings but, in general, a limited amount of machining would be needed on penetrator component surfaces because of their relatively large size.

5.2.7 Steel defects

Any steel contains defects such as slag inclusions and small pores or pockets of gas. Their size and frequency can be reduced to very low levels by longer and more sophisticated refining processes but the corresponding additional costs soon become prohibitive if unrealistically severe standards are specified.

The selection of appropriate defect acceptance levels for penetrators is thus very important because of its bearing on the choice and costs of steelmaking and component fabrication routes.

A major defect would seriously affect the structural and corrosion performance of a penetrator. It could cause a stress concentration and lead to crack propogation and eventual splitting failure in a penetrator under handling, pressure or thermally induced loads. The possibility of defect induced structural failures would have to be evaluated for both assembly, transport and other loads before seabed embedment, as well as hydrostatic pressure and thermally induced stresses after embedment.

The effect of defects on corrosion performance would generally only be a long term consideration for penetrator behaviour in the seabed. A gross defect extending over a considerable proportion of a penetrator wall thickness could clearly lead to early breaching of the engineered barrier and defects in general could be possible preferential areas for local corrosion.

A preliminary evaluation, however, suggests that very strict limits on acceptable defect sizes will not be necessary. Loadings on the penetrators before disposal will be relatively minor compared with the structural capacity of the thickness of steel provided to give a sufficient corrosion life. Cyclic loadings which can cause fatigue failures will be negligible and, most significantly, the major stresses induced in penetrators, principally by hydrostatic pressure, will be compressive rather than tensile.

Corrosion initiation at defects is also not seen as unduly significant. Data on local corrosion mechanisms such as pitting have been obtained from experiments on samples which contain characteristic steel defects.

At this stage, therefore, it is not necessary to specify unusually severe standards for steel quality. As a result, it is proposed that defect acceptance levels typical of those currently applied for high integrity engineering applications should be adopted as guidelines for penetrator construction.

Certain surface defects could be repaired by grinding but repair by welding would be discouraged as this could be more detrimental than the original defect.

A selection of defect acceptance specifications currently in use is given in Appendix C.

5.3 Reference Penetrator Tube Dimensions

The length of penetrator tubes has been made equal to the stacking height of the HLW canisters plus nominal 50mm spacings between adjacent canisters for facilitating complete penetrator void filling.

The 500mm internal diameter of the penetrators would leave a nominal 35mm annulus around the canister to allow void filling.

5.4 Design Variations

During the development of the reference penetrator design a number of interesting possible alternative details were considered. A selection of these is presented in Appendix D.

5.5 Conclusions

The development of a manufacturing standard for the reference penetrator design has been described in this section. The specification formed a basis on which discussions could be held with manufacturers and researchers to assess the technical considerations and approximate costs of fabricating penetrators. The results of these discussions are reported in Sections 6 and 9.

The main features of the specification developed for the steel bodied reference penetrator can be summarised as follows:

a) penetrator construction requirements appropriate to current manufacturing procedures have been selected.

b) a weldable, high purity 0.15% carbon mild steel with routine mechanical properties has been selected for penetrator components.

c) manufacturing tolerance limits and steel defect acceptance levels based on pressure vessel design requirements have been assumed to be appropriate at this stage of penetrator engineering studies.

6.0 PENETRATOR MANUFACTURE

6.1 Introduction

Penetrator manufacture has been considered as having two distinct stages namely:

a) component fabrication in conventional steelworks

b) assembly of components by inserting waste canisters and sealing the penetrator under shielded conditions in a special plant

Fabrication of large steel sections is a well developed industry and there are many candidate processes which could be used to manufacture the penetrator body, and the nose, tail and instrument hold sections.

Whatever fabrication technique is adopted, at least one circumferential sealing weld will be needed and so welding techniques for thick sections are reviewed.

Although minimising the amount of welding may be viewed as a desirable objective, restricting welding to one final weld precludes the use of some promising tube manufacturing techniques. To ease assembly and avoid the problems of remotely welding the tail to the body, it is suggested that this weld is performed during fabrication, leaving the nose to body weld to be made during assembly. In the report these are referred to as fabrication welding and assembly welding respectively.

As has been stated earlier, the expected corrosion rates for penetrators cannot be confidently predicted and the provision of a 75mm thick mild steel shell may be too much or too little. Penetrator tube manufacturing techniques have therefore been studied for a variety of wall thicknesses, namely 25, 75 and 200mm and also for lengths appropriate to single, five and twelve canister options.

A conceptual assembly plant design has been developed to the level at which it has been possible to verify that assembly operations are feasible and to enable basic costings to be determined.

6.2 General Fabrication Considerations

6.2.1 Steel specification

Ease of welding was the principal criterion for the development of the guide for steel chemical composition and mechanical properties described in 5.2. It was judged that the most appropriate specification was one based on current general engineering low carbon steel grades with additional requirements for high purity. The proposed steel is, therefore, not only readily weldable but would also be suitable for fabrication by any of the processes discussed later in this section. Discussions were held with several major manufacturers during the course of the study and none expressed any reservations about the compatibility of the steel specification with their production methods.

Only two general and well known points will be recorded here to illustrate that fabrication methods are not particularly sensitive to varying steel composition over a wide range:-

a) low carbon steels such as that described in 5.2.1 are readily forged because of their relatively modest strength and good ductility properties. Modestly increased forging complexity is only considered to be introduced for steels with more than 0.65% carbon or more than 5% total alloy element content by weight (Ref. 36).

b) Casting of steels is in fact considerably more difficult than for cast irons because of their higher melting points and lower viscosities. For example, a 0.15% carbon steel has a melting point of about 1550°C compared with 1150°C for a 4.5% carbon cast iron. Even so, similar processes are used for steel castings as for iron castings. The main differences are cost and tolerances in that steel refining and casting can be twice as expensive per tonne and steel castings are more liable to cooling distortions than iron castings.

6.2.2 Steelmaking capability

Most of the aspects to be considered during steelmaking, namely furnace and refining process selection, killing, grain size refinement and vacuum degassing were discussed in 5.2. The associated topic of heat treatments to achieve specified mechanical properties was reviewed in 5.2.3. There are, however, a host of factors, not all strictly technical points, that dictate whether suitably experienced high quality manufacturers would be in a position to quote competitively for penetrator construction contracts. One such aspect, minimum batch size, is outlined below.

There is invariably a minimum economic batch size for the steel melting facilities installed by a particular manufacturer. These would not be expected to influence the design of penetrators but might restrict the number of manufacturers who would be in a position to fabricate components for them. The implications of this would have to be considered at any detailed planning stage to ensure that sufficient companies with the necessary experience and facilities for high quality products would be able to fabricate the components for proposed penetrators.

This raises the possibility of one manufacturer being retained to supply relatively heavy items such as large penetrator tubes whilst another supplied the lighter nose and tail components. Factors such as slight incompatibilities between the steels used by different manufacturers would then have to be considered.

6.2.3 Quality assurance and quality control

The importance of quality assurance and quality control, QAQC, systems operated by manufacturers and clients is now generally recognised. It is the art of minimising defective production so that the disruption and expense of rectification or even component scrapping is avoided.

QAQC systems have three main stages being:-

a) the identification and definition of the necessary quality standards for a product to have a satisfactory performance.

b) the establishment of appropriate production procedures to achieve the defined quality standards.

c) the verification that the defined quality standards have been attained in the products.

The initial definition stage requires that the function of a product is first specified to enable required quality standards to be properly determined. A thorough analysis of this kind is valuable for establishing which standards would be sensible and necessary as opposed to those which would be unrealistic and of no real purpose. The most logical and economic approach to specifying the product should, therefore, be found in this way. The resulting quality standard requirements then have to be fully defined and recorded for the information of manufacturers.

The second stage requires the manufacturer to propose appropriate production procedures and prove, to the satisfaction of the purchaser, that they would give the agreed levels of product quality. Comprehensive testing of prototypes from trial production runs is necessary in some cases to establish the suitability of a proposed procedure.

During the proving phase it is vital that all production parameters such as temperatures and timings are noted in a process record because all steps during subsequent production have to be monitored and compared with the approved record. Any variation from approved parameters becomes a cause for the special examination of the production batch concerned. This is perhaps the most important part of the verification stage of a QAQC system in that products are accepted if the parameters monitored during their manufacture match the approved record.

Techniques such as the non destructive examination (NDE) methods need only be used on samples at preset intervals to ensure unexpected production variations are not occurring and on any batches affected by production parameter deviations. This concept of directed NDE is more economic and can be more effective than blanket NDE of all production and is discussed further in 6.7. Table 6.4 indicates the relative importance of QAQC controls at various stages of a fabrication sequence.

Appropriate QAQC systems for penetrator construction could be based on one of several guidelines which have been published in recent years for both nuclear (Refs. 37-39) and non nuclear engineering applications (Ref. 40). In this regard it should be mentioned that the nuclear power plant QAQC recommendations of British Standard 5882 are intentionally similar to those of the International Standards Organisation and the International Atomic Energy Authority.

6.3 Tube Manufacture

Seven processes for forming penetrator body tubes have been investigated and several of them have been shown to be suitable and well within the capability of present day industry.

The methods are described in the following sections only in sufficient detail to illustrate their feasibility as penetrator body manufacturing processes. They are discussed more fully in Appendix E which gives a greater appreciation of their range of applicability and the degree of specialisation involved in each.

6.3.1 Tube forging and machining

This process is shown on Figure 16. A large ingot is cast using vacuum degassing and the ends are discarded. The squat ingot is then forged using an open die process until it has been elongated to the required length and diameter. The ends are again discarded, the bore is cut using a trepanning tool and the outer surface is machined down to the finished size.

6.3.2 Tube solid casting and machining

Figure 17 shows the hollow sand casting technique. Solid casting takes place without the sand core. The molten steel is poured and allowed to cool before the mould is struck and the casting removed. Following heat treatment and finishing the central core is cut out using a trepanning tool.

6.3.3 Tube hollow casting

The process is shown on Figure 17. It is similar to the solid casting method except a sand core is placed in the mould and the molten steel is cast as previously. The sand core is removed when the casting has cooled and solidified and the inner and outer tube surfaces are machined to finished sizes.

6.3.4 Tube centrifugal casting

The process is shown on Figure 18. An external cylindrical mould is supported horizontally by driven and idling rollers near the ends. The rollers spin the mould and when molten steel is introduced the centrifugal forces form the tube. High accelerations are achieved and these ensure a good quality steel where any impurities are forced to the inner surface and can be removed by subsequent machining.

6.3.5 Tube extrusion

The process is shown on Figure 19. A large ingot is heated and upset once to remove scale. It is then placed in a blocking pot and made into a squat cylinder. A piercing punch is forced down the middle of the block of steel to form a central hole. The hole through the block is then completed using another punch and the ingot is placed in a large press where by placing an internal punch into the hole and forcing an extrusion head down over the ingot, the metal is caused to flow to form an extruded thinner tube which has both the correct inner and outer diameters.

6.3.6 Other processes

Plate forming and seam welding, as shown in Figure 20, is only possible for thin walled penetrator tubes up to 40mm thick. Pipe milling is also a possible process but it is only economical for large pipe productions runs.

6.3.7 Tube machining

Machining is a prerequisite for forged tubes, almost inevitable for large cast tubes and possibly required, though to a lesser extent, for other fabrication methods to meet the guide penetrator tolerance and surface texture specifications described in Appendix C. The fabrication options which, of necessity, entail extensive machining to produce plain tubes have the advantage that only modest additional processing complexity is involved for tubes with integral nose machining. This applies to the rough forging and solid casting procedures described in 6.3.1 and 6.3.2.

Typical details of bore machining for open tubes and blind-end tubes are given on Figure 21.

6.3.8 Tube with integral nose fabrication

The fabrication of penetrator tubes with integral noses to minimise welding is a cumbersome undertaking which can only be achieved with complete confidence by procedures involving extensive and expensive machining. Consequently, it is doubtful that the elimination of a weld between separately fabricated tubes and noses could be justified on purely economic grounds.

Inevitably, one circumferential weld would be required for penetrator closure and arguably there is little benefit in avoiding a second. This view is reinforced by the fact that the second circumferential weld could be made at the component fabrication stage and would therefore be operationally simpler to make. At the component fabrication stage, any heat treatment required could be provided without concern for damage to waste or filler material.

Therefore, the component fabrication assessment has concentrated on contrasting the options for manufacturing plain tubes on the basis that separate manufacture and welding of noses and tails would be most appropriate.

For completeness, however, a description of the machining processes to fabricate tubes with integral noses from forgings or solid castings is given in Appendix E. The possibility of hollow casting tubes with integral noses by using chaplets is mentioned in Appendix E although it is not seen as a particularly suitable solution. Extrusion of tubes with one end closed to form an integral nose is also feasible but is considered to be a most unlikely candidate because of production complications and size limitations and so has not been recommended.

It is probably fair to say that the increasing adoption of mathematical modelling and computer analysis methods is transforming ultrasonic testing just as radically as the control over arc welding parameters is being improved by the introduction of electronic operating units as discussed in Appendix F. The rapid advances hoped for in NDE technology, therefore, means that a detailed review of existing practices would merely be of transient relevance to the development of likely penetrator manufacturing procedures. Consequently, only a summary of current industrial and experimental NDE capabilities is given to show that they are consistent with the guide defect acceptance levels proposed for penetrators.

It is worth emphasising that NDE is a relatively small component of QAQC during a fabrication sequence as evident from the example given in Table 6.4 of a check point allocation to the various stages in the construction of a pressure vessel. It should also be noted that most NDE has conventionally been directed at checking critical welds with only intermittent sample inspection of other welds and parent material. A similar approach would probably be justified for a penetrator construction programme.

An interesting extension of NDE and in service monitoring is that the data collected can allow realistic assessments to be made of the probability of failure of fabrications. This could be of relevance to detailed radiological assessments of a penetrator disposal operation as a frequency could be estimated for the premature breaching of the engineered barrier containment of the HLW. A current example of the method is its application to evaluations of pressure vessel reliability (Ref. 44).

6.7.1 Molten steel sampling

Ladle or cast analyses of molten steel before it is poured into forging ingots or casting moulds is not considered as NDE but is mentioned here to demonstrate current practice for checking compliance with steel composition requirements. Various British Standards set out appropriate frequencies of sampling of refined molten steel for different end uses such as rolled, forged or cast products. Sophisticated spectrographic equipment is already installed by manufacturers of high integrity specialist products to give fast and accurate chemical composition analyses. Therefore, it would be reasonable and practicable to expect detailed monitoring of the composition of steels used in the manufacture of any penetrators.

6.7.2 Internal defect detection

Ultrasonic inspection has superseded radiography as the predominant method of internal defect detection. When contrasted with ultrasonic inspection, radiography is generally considered to be more certain for detecting volume defects like gas porosity or slag inclusions but less proficient at identifying crack like defects such as lack of fusion in welds. The safety problems associated with radiography and the recognition that crack like defects are usually most critical for initiating engineering failures has favoured the increasing adoption of ultrasonics.

Radiography is still used extensively, however, for checking welds in austenitic stainless steels which have an unfavourable metallurgical grain structure for ultrasonic testing. Radiography is also used if a permanent record, that is photograph, of the weld test is required as methods for magnetic tape recording of ultrasonic tests are only at an early research stage. Therefore, the sensitivity of existing radiographic techniques is briefly outlined in the next section although most attention is given to ultrasonic methods.

Eddy current, residual stress and acoustic emission are other detection methods that have evolved and some could justify further development as electronic equipment becomes more sophisticated. As yet, however, they are not the subject of major research programmes to the same extent as ultrasonics and so have not been evaluated in this study.

6.7.3 Radiography

Isotope sources such as cobalt-60 can penetrate up to 200mm thick steel but very long exposure times are needed so high energy accelerators, if available, are favoured for moderate to thick steel section radiography. Extremely thick steel can only be radiographed using very powerful accelerators and installations exist for inspecting greater than 500mm steel sections. As a general rule the cost of radiographic equipment and accompanying shielding facilities escalates rapidly if they are designed to inspect greater than 100mm steel thicknesses.

There is a remote possibility that radiography could be used for the NDE of assembled penetrators by virtue of the radiation emitted by their HLW contents. The relatively high gamma energy output of caesium-137 in up to 100 year old HLW would probably allow radiography of about 100mm thick steel bodies, without making allowance for the attenuating effect of any penetrator void filling materials.

Sensitivity of radiography has a nominal range of between 0.5 and 4% but 1.5% is a typical realistic target. Thus defects with a dimension exceeding 3mm should be detected on radiographs of 200mm steel thicknesses. This would be sufficient to check that the guide defect acceptance levels discussed in 5.2.7 had been satisfied in penetrators although a limitation of radiography is its relative inability to identify certain orientations of crack like flaws.

6.7.4 Ultrasonic inspection development programmes

There are currently intensive efforts being made in several countries to refine existing and develop more advanced ultrasonic inspection techniques for industrial applications. The requirements for very reliable and sensitive NDE of nuclear reactor components has been the main catalyst for this work. In France, for example, the development of focused probes has been consolidated for ultrasonic inspection of thick steel components for a major nuclear reactor building programme. Similarly, the possibility of approval being granted for construction of a PWR has stimulated the further investigation in Britain of ultrasonic techniques.

Several methods with varying potential for development have been evolved. Examples are -

(a) pulse-echo
(b) ultrasonic holography
(c) ultrasonic diffraction (TOFT)
(d) ultrasonic scatter
(e) ultrasonic surface wave mode conversion
(f) ultrasonic surface wave inspection
(g) ultrasonic spectroscopy
(h) crack tip reflection

Only brief reviews of methods a, b and c for penetrator construction are included in this section. No details of the theory and practice of the methods are given as these are reported elsewhere (Refs. 45 and 46).

(a) Pulse-echo: this is now the most widely used technique in industry having superseded earlier obscuration or decibel drop systems. It remains, at present, as a manual testing method and is consequently very dependent on the skill of operators for successful application. A 90% crack detection rate has been reported as a typical level of reliability. Another limitation is the need for focused probes to avoid excessive attenuation when inspecting thick walled components. However, ±2mm crack sizing accuracy can be attained in suitable conditions although this depends on the size of weld slag inclusions which can affect the sensitivity of the method.

(b) Ultrasonic holography: this is a promising technique because its lateral resolution is relatively unaffected by the thickness of test pieces and it gives permanent inspection records in the form of magnetic data tapes which can be analysed to give three dimensional size estimates of defects. Potential accuracies of ±2.5mm lateral and ±7.5mm through thickness sizing of defects have been reported.

(c) Ultrasonic diffraction: The Time of Flight Technique (TOFT) is being developed for possible automatic NDE of up to 250mm steel thickness components. It is not significantly sensitive to surface roughness. However, it cannot be used to detect defects within about 25mm of the near face and within about 5mm of the far face. Therefore, it has to be used in combination with other methods such as pulse-echo to check the full depth of a test piece. The aim is to locate defects with a dimension greater than 2mm at data acquisition speeds of around 2 to 4 m/hr for weld inspections. Minimum defect size detection levels of current experimental equipment is between 4 and 13mm depending on depth.

6.7.5 Current ultrasonic inspection sensitivity

The previous section outlined current ultrasonic inspection research to indicate the trend towards automation and techniques that record data on magnetic tapes which can be retained as test records. Thus developments are in hand to overcome the drawbacks of existing ultrasonic methods so that significantly improved procedures would be available for any future HLW penetrator disposal programme.

However, the sensitivity of present industrial practices rather than that expected of new systems under development has been investigated for the purposes of this engineering feasibility study. In general, ultrasonic inspection has similar sensitivity to radiography. Thus defects, particularly crack-like flaws, with a dimension exceeding 3mm should be detected and this would be sufficient to check whether the guide acceptance levels had been satisfied in penetrators.

6.7.6 Surface defect detection

Dye penetrant and magnetic particle techniques are well tried surface defect detection methods which could be readily applied to the outside of penetrators. Checking for defects on the inward facing surfaces of penetrator components would be less convenient but still possible by using video or fibre optic viewing equipment with consequent additional NDE costs. However, current ultrasonic internal defect detection techniques using probes on the outside of penetrators could locate flaws which are more than about 5mm from the internal surfaces. Therefore, it could prove more practicable to specify defect size limits for internal surfaces so that only ultrasonic testing would be necessary to detect unacceptable defects.

6.7.7 Directed NDE

One target currently under investigation at the Welding Institute is the integration of fabrication process monitoring and NDE procedures. For example, the increasing sophistication in the control and recording of welding parameters may lead to the adoption of directed NDE as opposed to time consuming blanket NDE of high integrity joints. That is, NDE would be concentrated in areas where a weld parameter fluctuated outside its normal range such that the creation of a defect might be suspected.

6.8 Assembly Plant Design

One purpose of this study was the estimation of likely orders of cost, as reported in 9.0, for an HLW penetrator disposal programme. Consequently, it was necessary to produce the plant design shown in Figures 28 to 33 so that approximate capital and operating costs of facilities for assembling penetrators could be assessed. It is stressed that the design should not be seen as a detailed proposal because the limited objective of approximate costing did not require a particularly accurate building scheme to be developed. Another consideration is that the commissioning of any penetrator disposal operation would be so far in the future as to make a fully evolved plant design of only passing value because of the changes in engineering practices which are bound to occur.

6.8.1 Main criteria

The manner in which the penetrator assembly plant design evolved was mainly dictated by the following criteria for its operation.

a) Reference penetrator design: the plant is arranged for the assembly of the five canister capacity 75mm wall thickness reference design of penetrator as shown in Figure 15.

b) An annual penetrator assembly rate of 115: the plant is sized for handling $85m^3$ of HLW annually.

c) Plant proportions: the plant includes facilities such as operating areas, stores and workshops in addition to shielded working corridors for penetrator assembly. The selection and relative proportions of these facilities are based on current practices for the design of nuclear installations.

d) Day shift plant operation: the plant is designed assuming as many operations as possible would be undertaken during day shift working hours. A nominal 2000 hours per year normal working time for penetrator assembly is assumed.

e) Horizontal penetrator assembly: the plant is arranged for assembling penetrators in a horizontal or near horizontal attitude. This is an important criterion because a very different plant layout would be required for vertical handling of penetrators.

f) Assembly equipment selection: the plant is designed assuming the installation of automated versions of current conventional industrial equipment. This is considered appropriate at this stage given that it leads to conservative estimates of penetrator assembly cycle times.

6.8.2 General considerations

The specific arrangement and sizing of facilities within the plant resulted from general considerations of the design and equipment of nuclear installations as follows:

a) Crane lifting ranges: there are considerable difficulties in designing electrical and mechanical systems for cranes to have particularly sensitive handling characteristics over a wide load lifting range. Consequently, the plant is arranged along three axes corresponding to distinct overhead travelling crane, OTC, service load rating requirements thus:-

OTC1 - nominal 120t rating for handling nuclear transport flasks in the flask handling bay.

OTC2 - nominal 5t rating for handling HLW canisters in the canister preparation corridor.

OTC3 - nominal 20t rating for handling assembled penetrators in the penetrator assembly corridor.

A secondary advantage of segregating different lifting operations in this way is that it results in appropriate spans for the different cranes. It should be noted that OTC1, 2, and 3 refer in effect to crane tracks which could carry more than one crane set if necessary for plant operations. For example, two crane sets could be installed on track OTC1 for simultaneous handling of canister and penetrator transport flasks.

b) Mechanical equipment minimisation: reliance is placed on overhead travelling cranes for handling movements inside the shielded working corridors to minimise the need for mechanical equipment and their associated maintenance problems.

c) Canister and penetrator storage: the size of the plant is most dependent on the amount of storage capacity provided to cope with fluctuations in the supply of canisters and the dispatch of assembled penetrators. Nominal capacity for 40 weeks supply of canisters and 20 weeks production of penetrators has been incorporated in the plant design.

d) Shielding of working corridors: the plant design includes provision of 1.5m nominal thickness concrete shielding for attenuating radiation from HLW in the canister preparation and penetrator assembly corridors. Access to the working corridors would be via lead lined shielding hatches and doors as shown schematically on the Figures of the assembly plant. It is possible that additional doors would be provided at intermediate positions along the corridors of an actual plant to enable maintenance staff to enter an area without requiring HLW to be removed from the entire corridor. An example of such a door is shown between the breakdown cell and canister preparation corridor.

e) Building construction: the Figures giving the plant design indicate a clad structural steel frame building shell enclosing reinforced concrete shielded working corridors. This is in accordance with current construction practice whereby the steel building shell is erected at an early stage so that the majority of reinforced concrete work can take place under cover from the weather.

6.9 Plant Operation

This section outlines a possible operation sequence for the penetrator assembly plant. It is acknowledged that many variations to the single procedure given here could be devised. Assembly processes are particularly sensitive to the approach adopted for the design of penetrators. Other stages, for example transport, depend on factors such as availability of rail access and the flask design practices in force at the time of plant construction. The plant operation sequence has, therefore, been developed for the assembly of the reference design of penetrator assuming the use of current nuclear industry practices wherever applicable. Certain checking stages and staggered shielding door opening requirements have been omitted from the sequence descriptions for brevity.

6.9.1 General remarks

Some general aspects of plant working procedures have been omitted from subsequent operation sequence descriptions to save excessive repetition. The following examples are included in this category of procedures:

a) Inspection steps: all aspects of the canister preparation and penetrator assembly process would be closely monitored and checked at appropriate stages. Examples of inspections are transport flasks for impact damage, canisters for incomplete ullage filling and penetrators for defective welding and void filling.

b) Shielding door opening sequence: the pairs of sliding shielding doors giving access to shielded working corridors would be operated on the air lock sequence principle. Thus a door would not be opened unless the other door in the pair was closed to maintain the shielding. Air handling equipment would maintain appropriate pressure variations in the plant to ensure flows were towards rather than away from shielded working areas.

c) Decontamination procedures: no item, including travelling cranes, would be moved out of the shielded working corridors before being checked for and, if necessary, cleaned to remove radioactive deposits. Decontamination areas for this purpose are indicated on the assembly plant design Figures.

d) Miscellaneous component and material handling: the handling of items such as penetrator noses and welding consumables is not detailed in operation sequences as it is secondary to the main functions being described.

6.9.2 HLW canister supply

It has been assumed that HLW canisters would be brought from a vitrification plant storage installation by rail in transport flasks weighing roughly 120t. A flask would hold about 20 canisters each containing 0.15m³ of HLW. Therefore, 29 flask movements entailing a minimum of two to three train loads would be necessary to deliver a year's supply of canisters totalling 85m³ of HLW.

A typical canister supply sequence following the arrival of a train of flasks at the assembly plant is shown in Figure 32 as follows:

a) canister transport flask on railway wagon shunted into flask handling bay.

b) canister flasking-in cell hinged hatch cover opened.

c) transport flask lifted by OTC1 from wagon onto trolley in flasking-in cell.

d) transport flask lid bolts removed.

e) flasking-in cell hinged hatch cover closed.

f) trolley operated to move flask to unloading position of flasking-in cell.

g) flasking-in cell hatch cover opened by OTC2 and placed in adjacent parking area.

h) flask lid removed by OTC2 and placed on top of hatch cover in parking area.

i) canisters lifted out of flask individually by OTC2 and deposited in convection cooled storage bank.

j) when emptied, transport flask returned to railway wagon by reverse sequence of steps a to h.

6.9.3 HLW canister preparation

It would be necessary to fill the ullages of the HLW canisters for disposal using the reference design of penetrator. At this stage it has been assumed that lead would be used for filling.

This operation has been described as canister preparation. A typical sequence is shown in Figure 32 as follows:-

a) canisters lifted out of storage bank individually by OTC2 and lowered into a canister preparation station.

b) canister lid removed or punctured to give access to ullage.

c) ullage heated and surfaces tinned.

d) air evacuated whilst molten lead injected into the still hot ullage.

e) canister vibrated if necessary to ensure no voids become trapped in the lead during filling.

f) canister lifted out of preparation station and deposited in convection cooled prepared canister storage bank.

6.9.4 Penetrator assembly

A scheme for penetrator assembly is set out in Figure 34. At this stage it has been assumed that lead void filling and automated conventional arc welding processes would be used for penetrator completion as shown in Figure 35. Alternative welding procedures are shown in Figures 36 and 37 to illustrate the range of options for penetrator completion.

A typical assembly sequence is shown in Figure 33 apart from prepared canister transfer which appears on Figure 32. The steps in the assembly sequence would be:-

a) penetrator components on road or rail wagon moved into unloading bay.

b) components unloaded by OTC3 and deposited in penetrator body storage racks.

c) when required, components taken from storage racks by OTC3 and moved to penetrator loading station in line with the transfer tunnel.

d) canister lifted out of prepared canister storage bank in canister preparation corridor by OTC2 and placed on transfer tunnel trolley.

e) trolley operated to move canister into transfer tunnel.

f) equipment in transfer tunnel used to rotate canister from vertical to horizontal attitude and to insert it within a frame, see Figure 34, for locating canisters in the penetrator body.

g) steps d to f repeated until frame completed by insertion of fifth canister.

h) trolley operated to move frame through transfer tunnel to the penetrator assembly corridor.

i) frame with canisters loaded into waiting penetrator body.

j) penetrator moved by OTC3 from loading station to a completion station where it would initially be held at a slight angle to the horizontal for void filling.

k) penetrator and canister contents heated then air evacuated whilst molten lead injected to fill voids between the penetrator body and canisters.

l) penetrator vibrated if necessary to ensure no voids become trapped in the lead during filling.

m) penetrator position in completion station adjusted using OTC3 to be in horizontal attitude for nose welding.

n) penetrator nose welded to body, penetrator repositioned in inclined attitude, nose void heated then filled with lead before nose filling hole sealed by weld deposited metal as shown in Figure 35.

o) any necessary residual stress relieving undertaken to complete penetrator.

p) penetrator lifted out of completion station by OTC3 into convection cooled assembled penetrator storage racks.

6.9.5 Penetrator dispatch

The penetrator dispatch arrangements included in the assembly plant design assume similar transport criteria to those described for canister supply. Thus a procedure is outlined in which assembled penetrators are loaded into transport flasks for dispatch by rail to the disposal ship loading dock. It is estimated that the total weight of an assembled reference design penetrator carried in a flask meeting the requirements of the current IAEA Transport Regulations would be about 120t.

The penetrator dispatch arrangements adopted in any future proposals could take more account of compatibility with disposal ship loading procedures. Radically different penetrator transfer methods would, in particular, be envisaged if the assembly plant and disposal ship dock shared the same site. Several examples of penetrator dispatch schemes are discussed in 7.0.

The typical penetrator dispatch sequence given below is shown on Figure 33.

a) empty penetrator transport flask on railway wagon shunted into flask handling bay.

b) penetrator flasking-out cell hinged hatch cover opened.

c) transport flask lifted by OTC1 from wagon onto trolley in flasking-out cell.

d) transport flask lid bolts removed.

e) flasking-out cell hinged hatch cover closed.

f) trolley operated to move flask to loading position of flasking-out cell.

g) flasking-out cell hatch cover opened by OTC3 and placed in adjacent parking area.

h) flask lid removed by OTC3 and placed on top of hatch cover in parking area.

i) penetrator lifted out of assembled penetrator storage racks by OTC3 and lowered into waiting flask

j) when loaded, transport flask closed and returned to railway wagon by reverse sequence of steps a to h.

6.9.6 Canister and penetrator rectification

Facilities for rectifying any faults that occurred during canister preparation or penetrator assembly are indicated in several locations on the Figures showing the assembly plant design. An illustration of their purpose would be if an unwanted void was detected in the lead filling a canister ullage. The canister would be taken to a rectification station where, for example, the lead might be remelted and the void eliminated.

In extreme circumstances it might be necessary to completely dismantle a penetrator. Areas are allocated in the plant design for equipment to cut up assembled penetrators and retrieve HLW canisters. The debris of the penetrator would then be disposed of separately.

6.9.7 Plant maintenance

It is envisaged that the plant would be closed down and cleaned to allow staff to enter and service equipment in the shielded working corridors about once a year. Routine maintenance during plant operation would be undertaken by remote control using robot attachments on overhead travelling crane sets. Equipment would be of modular construction to facilitate repair and replacement in this way.

If necessary defective equipment could be taken by OTC2 into the dismantling cell at the end of the canister preparation corridor, Figure 32. Equipment from the penetrator assembly corridor would be moved through the transfer tunnel when being taken to the dismantling cell. Remote controlled manipulators in the dismantling cell would then be used to repair the equipment or dismantle and pack it into drums for disposal as low or medium level radioactive wastes as appropriate.

The decontamination facilities included between the shielding doors at one end of the breakdown cell could also, if necessary, be used to allow equipment from the shielded working corridors to be taken into the controlled workshop for direct manual maintenance by servicing staff.

6.9.8 Ancillary facilities

The provision and operation of ancillary facilities necessary in nuclear installations has not been considered in detail as this would be an academic exercise at this early feasibility stage. Air venting and filtering equipment, electrical switchgear, cooling water supply, consumable material feeds to items such as welding units in shielded working areas, health physics and changing rooms for staff, low level radioactive waste management procedures, offices and computers are all examples of the wide range of facilities that would be installed in the operating and service areas and administration block.

6.10 Plant Capacity

It is not considered that there would be significant cost differences between alternative plant designs for a particular penetrator production rate. However, costs would clearly be greatly affected if a much larger plant or, perhaps, two or three similarly sized plants were built to suit a higher penetrator production rate. The factors limiting assembly plant output and the ramifications of designing for different disposal rates are, therefore, discussed here.

6.10.1 Plant output limitations

The main output capacity limitations for the assembled plant design shown in the Figures is the rate at which assembled penetrators could be dispatched via the flasking-out cell. There are typically 200 normal working days for an industrial operation during a year. An optimistic maximum is considered to be two flask loading cycles per day giving a limit of 400 penetrator movements per year at full utilisation. In practice, however, penetrator dispatch might have to be co-ordinated with disposal ship sailings so that it is unlikely that penetrator flasking out would be a continuous process. Average utilisation throughout a year of 40 to 60% is regarded as a realistic range at this stage giving a practical maximum annual output of about 240 penetrators equivalent to 180m^3 of HLW per year.

A secondary but significant consequence of increasing the design output to the peak throughput of the flasking-out cell would be the need for a corresponding increase in assembled penetrator storage capacity of the assembly plant. As a rough guide, it is considered that the total volume of the plant would have to be increased by 40% to accommodate a 100% increase in design output.

6.10.2 Plant output enhancement

There are two basic ways of enhancing the maximum output capacity of the assembly plant. Multiple shifts could be worked to increase utilisation or the design amended to allow more than one flask to be handled at once. At this stage it has been decided not to assume more than day shift working of a plant leaving enlarged flasking-out facilities as an approach to improving output capacity. An enhancement of more than 50% is considered unrealistic for a single penetrator assembly corridor. This would give an absolute production limit of 360 penetrators containing 270m³ of HLW per year. The size of the assembly plant shown in the Figures would have to be increased by about 80% for this three-fold uprating of penetrator output.

Parallel penetrator assembly corridors added to the plant or more than one plant would be required for any greater penetrator production requirements. For example two penetrator assembly corridors arranged on either side of a central canister preparation corridor would be a suitable modification of the plant design concept shown in Figures 28 to 33. This solution would also involve an approximately 40% plant size increase for a 100% penetrator output enhancement.

6.10.3 Alternative capacity plants

To summarise the discussion of this section, the effect of alternative HLW disposal rates on the provision of penetrator assembly facilities would be as follows. The design shown in Figures 28 to 33 is described as the reference assembly plant in this schedule.

a) 85m³ HLW/yr: one reference assembly plant would be required.

b) 170m³/yr HLW: two reference assembly plants or one 40% larger uprated plant would be required.

c) 400 m³/yr HLW: five reference assembly plants or two 60% larger double penetrator assembly corridor uprated plants would be required.

6.11 Conclusions

The results of discussions with manufacturers and researchers regarding possible approaches employing current technology to the fabrication and assembly of steel penetrators have been reported in this section.

The study has concentrated on assessing the technical aspects of the fabrication of three basic components, namely a nose, tube, tail and fins from which a penetrator could be assembled by welding. Any decision to minimise welding by manufacturing penetrator tubes with integral noses or tails would rule out certain high quality processes which are only applicable to the manufacture of plain tubes.

The size and quality to which steel tubes can be made has been identified as being the manufacturing factor most likely to influence any detailed penetrator design development. Various processes have been evaluated for the production of 500mm internal diameter mild steel tubes with up to 200mm thick walls in lengths up to 16m. This range corresponds to the different tube sizes for the Figure 13 penetrator design options. Table 2.1 and Figure 22 give the tube size manufacturing limits of each process assessed during the study.

A comparative evaluation of the tube manufacturing is given in Table 6.2 and the following general conclusions have been drawn:

a) Forging and machining: this is a versatile high quality production option by which penetrator tubes of suitable lengths and wall thickness could be made. However, it is a time consuming process with high energy and manufacturing plant overhead costs. Open die forging and machining are, therefore, more appropriate for short production runs of large, complex components and would be expected to be prohibitively expensive for producing large numbers of penetrator tubes. See Appendix E and Figure 16.

b) Static sand casting: uniform wall thickness tubes are particularly difficult to cast without the formation of steel solidification defects. Measures such as solid casting or hollow casting with external padding and extensive machining are needed to overcome this problem. Therefore, although static casting should not be eliminated at this stage as a candidate process, it is the least immediately attractive option, as regards quality and cost for the production of penetrator tubes. See Appendix E and Figure 17.

c) Centrifugal casting and tube extrusion: it is difficult to differentiate between these two processes in terms of their suitability for the production of penetrator tubes. Both methods are especially appropriate for production runs of high quality tubes with unusual wall thickness to diameter ratios. However, the size of tubes that can be made is limited to about 12t by the capacity of current production equipment. This can be compared with the 7t weight of the 6.5m long 75mm wall thickness tube for the reference design of penetrator. Either of these processes would appear to be suitable for penetrator tube production. See Appendix E and Figures 18 and 19.

d) Plate forming and seam welding: this would be a particularly economic approach to penetrator tube fabrication. However, its adoption would be dependent on radiological and corrosion research programmes concluding that a penetrator steel wall thickness of under 40mm would be satisfactory. This is about the maximum plate thickness that can be formed into the nominal 500mm internal diameter tubes proposed for penetrators. See Appendix E and Figure 20.

Nose and tail component fabrication has not been found to be a significant factor affecting the feasibility of satisfying desired penetrator performance specifications. A general assessment of manufacturing options has concluded that sand casting would be most practical for the relatively large length to diameter ratio of 1.5:1 assumed in this study for the reference design of penetrator nose and tail shapes. Forging would only be appropriate for blunter nose and tail forms. It is considered that separately fabricated fins welded onto tail pieces would most facilitate penetrator construction. See 6.4, 6.5 and Figures 23 to 27.

A comparative evaluation is given in Table 6.3 of the options considered for circumferential welds between penetrator nose, tail and tube components. On current evidence it is felt that welding is most likely to satisfy penetrator performance requirements.

Two distinct conditions under which penetrator welding would be undertaken have been considered, namely:

a) fabrication welding in conventional steelworks

b) assembly welding by automated processes in a special plant where HLW would be handled.

The steel composition and range of penetrator wall thicknesses considered in this study are weldable by current industrial arc welding procedures. The general conclusions reached are:

a) current examples of arc welding including the comparatively new narrow gap methods for thick section joints should be considered for penetrator fabrication welding as it is not possible to say to what extent other processes will become available at conventional steelworks in the future.

b) novel processes currently under development, particularly electron beam and friction welding, should be considered as strong contenders alongside automated arc welding to meet the special requirements of penetrator assembly welding.

The various welding options are outlined in Appendix F and shown schematically in Figures F1 and F2. Possible penetrator completion schemes employing welding are shown in Figures 35 to 37. A design for an assembly plant where HLW canisters would be loaded into penetrators has been described in 6.8 and 6.9 and shown in Figures 28 to 33. The design is not a detailed proposal because the limited objective of assessing likely orders of cost did not require a fully developed building scheme to be devised. It is estimated that the size of plant shown would be suitable for an annual throughput of $85 m^3$ HLW corresponding to the UK rate of disposal or half the CEC lower rate of disposal as defined in Appendix A.

7.0 SHIP DESIGN AND PENETRATOR RELEASE

7.1 Introduction

Section 3 dealt with the generic aspects of the disposal operation in sufficient detail to conclude that the penetrator assembly should be land based and that the disposal ships should only transport and release completed penetrators. Adopting these conclusions, the operations reduce to transferring the assembled penetrators onto the ship, the voyage and the release. These aspects are treated in more detail in this section.

The relevant factors are discussed and five options are presented which demonstrate the kind of systems which could be developed. The discussion shows that the length of the penetrator continues to differentiate between systems and has a strong influence on their design.

7.2 Shore To Ship Transfer

7.2.1 General

The transfer from the penetrator assembly plant to the ship will take place most appropriately at a coastal dock which offers most of the facilities of a small to medium sized port such as Barrow or Sharpness in the UK. No exceptional handling equipment is envisaged but it is expected that a dedicated dock will be necessary to provide the anticipated level of security and public acceptability. Security is unlikely to be a major problem as there will be no fissile material on the site and the waste canisters are too large and hazardous to be removed without authorisation. The possible need for security systems to prevent sabotage, vandalism or to control demonstrations in the vicinity of the site would be no more onerous than for existing nuclear establishments.

The labour force required to operate the dock would be numbered in tens and not hundreds, and it is unlikely that recruitment would pose any real problems. At present the dockers at Barrow handling spent fuel shipments are conventional (non radiation) workers and are not required to wear dosimeters and the same regime has been assumed for penetrator shipments. The Barrow dock is not classed as a nuclear licensed site in UK law.

7.2.2 Deflasking unit

The main store for assembled penetrators is intended to be built adjacent to the penetrator assembly plant. If this is sited close to the dock and the penetrators are stored in a form whereby they can be loaded onto the ship without further work, then there will be no need for another store. This ideal situation may not be possible due to numerous procedural or operational problems and if this is the case a small store may be required at the dockside from which the penetrators can be dispatched to the ship.

A store would also be necessary if the penetrators are required to be transferred from rail or road transport flasks to seagoing flasks.

7.2.3 Use of existing facilities

The ocean disposal of HLW using penetrators would probably start at a small scale and gradually build up. Before the system has been developed and perfected, it is unlikely that a major new purpose built dock complex would be commissioned. It is much more probable that an existing dock will be hired and the necessary adaptions made to enable the loading and dispatch to be conducted. A precedent has been set by BNFL at Barrow where an existing dock was used until the volume of traffic plus the acquired experience enabled a purpose designed facility to be constructed. Using an existing port has manifest advantages as the construction of a new medium to deep water port requires a long lead up period and is not in keeping with the downturn in so many British ports.

Assuming that the prospective site has the required water depth and that the tidal range is satisfactory, the next most important aspect will be the availability of the land required for the assembly plant and the store. Most ports of the required size would have had a rail link into the site at some time in the past and this may have to be resurrected. Good road access is also essential but again it is likely that the port will have had to cater for heavy road vehicles during its history and so reasonable access is likely to exist.

The old mechanical handling equipment is not likely to be much use. Cargo handling has been considerably modernised over the past few years and the well known traditional type of quayside crane is unlikely to have the capacity for handling penetrators. Outside shipbuilding yards, normal cranage is limited to 40t which is the weight of the 40 ft ISO container so commonly used.

There would appear to be few technical restrictions on the lifting capacity of modern cranes and up to 500t is quite feasible. There are, however, considerable economic disadvantages in purchasing a high capacity crane which will only be used intermittently. At Barrow, BNFL used a hired barge mounted 200t crane to lift the spent fuel flasks until their new dock which contains a dedicated 150t crane was ready.

7.2.4 Civil and mechanical installations

The nature of the transfer of penetrators onto the ship is not one of high volume or high speed. Safety, on the other hand, is paramount and reliable fail-safe systems are required.

The options discussed later cover a wide range of possible penetrator loading systems. The simplest require little more than the installation of a vehicular loading ramp of the kind provided at many roll on, roll off terminals throughout the world. These have been developed to accommodate high tidal ranges and many ship types. Three types are available, those which are permanently attached to the ship, those which sit on a barge and, finally, those which are permanently attached to the dock.

At the opposite extreme, a rather sophisticated underwater conveyor has been devised which involves the construction of a tunnel starting at the dock and stretching out under the water to terminate under the moonpool of the ship. The possible advantages of using water as the shielding medium during transfer are outweighed by the difficulty of maintaining moving underwater equipment and by the consequences of system failure.

Although it can be argued that the reliability of a system can be improved by the incorporation of additional equipment and system duplication, an inherently unsafe operation cannot be transformed into a safe one. This is shown in the preference for low speeds and, especially, low lifts which do much to limit the dangers.

Some civil construction will be necessary. This may involve ancillary buildings and, in some options, the construction of shielded tunnels through which the unshielded penetrators can pass onto the ship. None of the civil work is considered to be major and the eventual reinstatement of the dock to its original condition should not be prevented by the envisaged works.

7.2.5 Radiation exposure

Transfer methods have been proposed which are safe and reliable and which will not result in unacceptable dose commitments. Either by the use of transfer flasks or shielded tunnels, the operating radiation level can be reduced to currently acceptable levels and even mechanical malfunction will not result in high dose levels.

7.3 Ship Design

7.3.1 General

This section sets out to identify the major factors which affect the design and operation of the ships used to transport and release the penetrators. In Section 3 it was concluded that the ships should not include any penetrator manufacture or assembly functions and that their role was best limited purely to transport and release. It was argued that the conversion of existing shipping was unlikely to be successful and that purpose built vessels would be required.

The features which need to be incorporated into such a vessel are discussed in the following section. Detailed ship design is clearly the work of naval architects and the scope of this study is only to reach a level where they could be briefed should it be decided to progress with ship design.

7.3.2 Hull construction

Existing spent fuel transport ships which have been purpose built to transport spent nuclear fuel from Japan, provide the initial model for the disposal ship designs in this study. They include several special features which are not found in conventional ships and which fundamentally affect their construction.

The most notable of these are the wide wing tanks surrounding the midship cargo area. These give enhanced beam collision protection around the hold and help prevent breaching. The tanks are normally dry when the ship is laden but are ballasted when running light to adjust the buoyancy. The wing tanks of disposal ships could be designed to be flooded when penetrators are onboard, thereby providing shielding.

In addition, the ships are constructed with double bottoms to give protection against damage due to stranding. The resulting space is often used to carry diesel fuel. Existing ships carry enough fuel for the return trip making the provision of fuel storage a very important aspect of their design.

In the case of the disposal ships a much smaller fuel storage capacity is likely to be needed and therefore a higher proportion of the deadweight is available for cargo.

The protected holds of spent fuel transport ships are separated from the crew and engine accommodation at the stern and the working area at the bow by 1m thick cofferdams which can be flooded to provide additional radiation shielding as well as acting as transverse bulkheads.

Almost all large ships have a transverse watertight bulkhead set back from the bow to protect the ship from foundering following a head-on collision. Other transverse and longitudinal bulkheads are provided to give the required degree of reserve buoyancy. The number and frequency of these depend on many factors and are determined during detailed design. Watertight doors are incorporated when the bulkheads, which normally extend up to the weather deck level, have to be used for access between compartments. The weather deck is the deck separating the hull from the ship's superstructure.

It has been shown that the penetrator body shell does not provide sufficient radiation shielding to achieve the current recommended dose levels. Unless each penetrator is carried in an individual flask, the necessary shielding will be incorporated into the structure of the ship.

Lastly, there do not appear to be any abnormal speed requirements for the hull and modern merchant shipping speeds are appropriate. The inclusion of a moonpool through the hull may increase the drag but this is unlikely to have a major influence on the power requirements. Ship speed is proportional to the cube root of engine power and so there are very good arguments for limiting the design speed.

7.3.3 Mechanical equipment

The probability of a ship carrying nuclear waste suffering loss of power or manoeuvrability at sea and being unable to reach port would be minimised by providing a high level of back-up equipment. Twin engines and twin propellers are customary and these also provide high ship manoeuvrability.

Propellers are often of the variable pitch type which, when combined with bow and stern thrusters, make the vessel highly manoeuvrable in restricted port areas and help to keep station at sea while being subjected to lateral currents and side winds.

The engine room is best situated near the stern as this limits the length of the prop shafts and avoids them having to pass along the length of the shielded hold. If the engine room cannot be located at the stern, the main engines can be arranged to drive electric generators and the power can be conducted to electric motors at the stern to drive the propellers.

The ancillary equipment such as back-up generators navigational and fire fighting equipment is also provided to an above average specification.

In contrast, the onboard equipment required to handle and release the assembled penetrators needs considerable development. It is not customary to move cargoes during a voyage and the available mechanical handling equipment is normally used only in port. The added complication of having to use remote handling reinforces the argument that much development is needed.

As far as can be established, there is little military experience of handling large heavy objects on board. When this is done, as with the movement of aircraft onboard carriers, it tends to be a "hands on" process often using small battery operated tractors to provide the propulsion.

Because of the lack of experience in the design of remote mechanical handling equipment, it has been necessary to evolve some potential methods. In all but one case discussed later, the equipment used to load and store the penetrators is reused in reverse to release them.

The advantages of hydraulic power make it the most likely power source. Its resistance to weather, to misuse and its excellent controlability make it preferable to mechanical or electrical power. In addition other advantages are the simplicity of the mechanisms located within the shielded, active area and the ability to site the more complicated power plant in a non active area.

The release method would probably involve inspection and monitoring of the penetrator before and immediately after release to confirm satisfactory emplacement.

From the calculation of the anticipated numbers of penetrators, the speed of release is not critical to the overall operation.

7.3.4 Cargo storage

The methods of storing the penetrators on board have a fundamental effect on the design of the ship. The advantages associated with vertical, longitudinal and transverse orientations are discussed later with individual ship designs.

The storage medium can be wet or dry. The obvious advantages of cooling and shielding provided by wet storage are overcome by the disadvantages of the resulting contamination. Both kinds are employed in the designs presented later.

Flexibility to handle penetrators of varying lengths and diameters could be important but at present it is unlikely that the circumstances leading to the preferred design would change dramatically.

7.3.5 Navigational equipment

Modern satellite navigational systems such as Navstar have a claimed accuracy of 10 to 30 metres. At present, the system is restricted to military use but it is expected that by the year 2000, similar civil equipment will be available and that this equipment would be installed in the disposal vessels.

The incorporation of variable pitch propellers, and bow and stern thrusters means that the ship would be able to hold station at the chosen release point.

7.4 Penetrator Release

It is expected that the loaded vessel would sail directly to the disposal site and not make any intermediate ports of call for bunkering, provisioning or to load additional penetrators.

The penetrator release mechanism will have been based on a maximum design sea state and it has been assumed that this sea state will only be exceeded for short spells during the release period.

It has been assumed that the disposal vessel would align itself relative to the prevailing direction of the waves to achieve the steadiest possible launch conditions. The envisaged size of disposal vessel would easily provide a stable launch platform for wave heights of less than 2m and from Figures 45 and 46 it may be seen that suitable sea states exist at the study areas for sufficient periods to easily accommodate disposal operations.

An appreciation of the sea and weather conditions at the study areas may be gained from Figures 45 to 49 which provide wave, wind and visibility data.

The ship will navigate to the predetermined location using its satellite system. Most probably it will approach slowly from down wind using the variable pitch propellers to arrive gently on station.

During the last stages of the manoeuvring when the ship is barely under way and the ships movements are at a minimum, the penetrator will be retrieved from store and conveyed to the release station. A final inspection will then be made to ensure that no damage has occurred during storage and transit.

At this time any instrumentation in the penetrator body will be activated and any safety lines removed.

The exact launch methods vary with the options and these are discussed for each ship design in the following sections.

7.5 Option A

7.5.1 Summary

This design uses the shielding properties of sea water to effect an underwater transfer from the land store into the ship using an underwater conveyor. A rail mounted ship gantry crane picks the penetrator off the conveyor and brings it on board through a stern moonpool. The penetrator is placed on a carriage running on the same rails and transported to the lateral storage racks, picked up and lowered into the store.

Release is a reverse of the loading operation except that when the penetrator is taken to the edge of the moonpool it is tipped in off the carriage.

7.5.2 Transfer

The assembled penetrators are loaded onto a conveyor in a shielded cell built adjacent to the quay. The penetrators sit transversely across the conveyor which starts in the cell above water, inclines down below water level and terminates beneath the stern moonpool of the disposal vessel. The penetrator is placed onto the conveyor by mechanical grabs and is moved until it is directly beneath the moonpool. This system could be used for the reference penetrator but the width of moonpool required would make it unattractive for longer penetrators.

A gantry crane onboard lowers its grabs down through the moon-pool, grips the horizontal penetrator and lifts up into the ship.

7.5.3 Ship

The ship design for Option A is shown in Figure 38. The engines and crew accommodation are located at the bow of the ship and separated from the shielded hold by a full width cofferdam. The main engines drive generators and the twin propellers are then driven by electric motors housed at the stern. This avoids the need to have long prop shafts passing through the shielded hold.

The thick sides and the deck above the hold provide shielding. They are unpenetrated apart from an access hatch in the top deck to service the gantry cranes. Cooling and ventilation are provided if necessary but the operation is remotely controlled and observed using CCTV.

The hold is subdivided by transverse bulkheads rising to the crane rail level. These make holds about 2 to 3m wide, each one of which can hold three horizontal penetrators stacked vertically. The width of the holds is the ships breadth minus the side tanks giving enough space for penetrators up to 10m long.

Hydraulic pistons can be actuated to close above a penetrator in the hold to provide the support for one to be placed above. The grabs on the gantry crane are positioned so there is no conflict with the vertical rack supports. The penetrator is lifted by three clamps, one near each end and one at the centre of gravity which acts as a safety clamp during normal handling. The clamps are hydraulically activated and grip the penetrator sides avoiding the need to have any lugs or protrusions on the body. This method was used successfully to handle very large naval shells until these became obsolete.

The ship and mechanical handling plant have the advantage of using well proven equipment, such as conveyors and gantry cranes, but there are also several disadvantages.

The need to have an underwater conveyor housing will require several metres of draught in addition to that needed for the ship itself. It will be difficult to maintain the conveyor mechanism and prevent it silting up if it is too close to the bottom or recessed in a trench.

Internally the hold subdivision makes the ship very strong and stable. The drawback is that the penetrators being loaded wet will then drip slightly contaminated water on the bottom of the hold. Unless this is washed out regularly the contamination will build up and cause problems. Washing systems exist but the added complication, expense and secondary waste production, are not attractive.

7.6 Option B

7.6.1 Summary

This design takes advantage of the shielding and cooling properties of water storage on board ship.

The short penetrators are stored in vertical racks on board and are released through gates built into the bottom of the ship's hull.

This option is not viable for the reference penetrator as the vertical storage arrangement and deck and hull gamma gate mechanisms restrict penetrator length to about 4.5m corresponding to 2 canister capacity.

7.6.2 Loading

Loading is effected by means of a shielded flask with a door set in the bottom. The penetrator is placed vertically up into the flask through the door in the base and the flask lifted by overhead crane onto the ship's deck and the penetrator lowered into its rack.

This transfer flask is designed to provide adequate shielding during transfer and would be strong enough to resist damage if it were accidentally dropped during the operations. The short timescale of the transfer should mean that no normal cooling would be needed and that emergency cooling could be provided by water hosing if this were necessary.

The door in the base of the flask would be a guillotine type similar to the gamma gate arrangements commonly used for on-site flasks at nuclear reprocessing plant.

As the penetrator is held by the tail a pintle arrangement is suggested for lifting and lowering using a hoist mounted within the transfer flask.

To minimise the hazard during transfer the flask would be lifted the minimum height needed to clear the side of the ship.

7.6.3 Ship and storage

The ship design for Option B is shown in Figure 39. Only one penetrator can be stored in each vertical rack and its length is restricted to the depth of water in the hold less the free space required to provide radiation shielding. To retain a reasonable draft penetrator length is restricted to about 4.5m corresponding to two canisters capacity, which would lead to a draft of about 7.5m.

The option B ship wet storage requires a deeper hull than for the other ship designs. This in turn requires the ship to be longer to retain reasonable proportions. The net result is that ship option B is larger than the other options.

The storage racks can be built up to weather deck level so the volume of water in the hold is well contained and should not shift during the voyage.

The weather deck is similar to the hull in that above each rack there is a gamma gate which mates with the base of the transfer flask. Any additional shielding can be provided by this deck.

The ship layout is conventional with the engines and accommodation aft. Wing tanks, floodable transverse bulkheads and a double bottom are standard features, although the bottom is highly perforated in this design and this weakens the strength of the hull. Not only is this a disadvantage but the area contains all the remote operated release doors which are in the least accessible place for maintenance and the most vulnerable place to be damaged in grounding accidents. The other main disadvantage is that the double hulled space cannot be used for fuel storage and that this will have to be provided elsewhere onboard.

7.6.4 Release

This is the only design which releases the penetrators in a nose down vertical position. When the ship is at the exact release station the bottom hull door will be opened and the hydraulic clamps securing the penetrator will be released and the penetrator allowed to fall into the sea. Because the fins protrude beyond the body section the release mechanism is complicated and guides will be needed to avoid the fins being damaged.

This method offers potentially, the simplest release operations requiring no transfer of the penetrator before the launch.

7.7 Option C

7.7.1 Summary

This option stores horizontal penetrators longitudinally in a dry hold. They are released through a centrally positioned moonpool. The loading and onboard transfer is effected by a rail mounted self propelled transfer flask.

7.7.2 Loading

This option is capable of handling the longest study penetrators containing 12 canisters (about 18m long). The weight of the penetrator and the shielding required during transfer mean that a crane with a capacity in excess of 200t would be required. In addition it is argued that for safety the penetrator should be raised the minimum distance above ground.

The shielded transfer flask is similar to a railway wagon with a bogie at each end. The bogies are mounted on central pivots so that they can be slewed to be at right angles to the body. This enables the flask to move laterally.

In the base of the flask are two pivoted doors which swing up inside the body to allow the penetrator to be hauled up into the flask. Three hydraulic clamps fit around the two extreme ends and the centre of gravity of the penetrator and it is raised into the flask using an internal hoist.

The base doors are closed and the flask is propelled from the land store by rail to the purpose built lift set into the quayside. In Figure 40 the lift is located to load over the ships side but if the disruption to the ships layout could be tolerated the design could be modified to load over the stern or bow.

When the lift has reached deck level a short connecting ramp is positioned and the flask is rolled onto the deck.

At this point the ship will roll due to the imbalanced weight at the deck edge although this could be minimised by careful ramp design. This problem is common to container ships which load transversely in this manner and the problem is counteracted by pumping ballast water between the wing tanks. This part of the loading may be slowed by the safe rate at which the water can be pumped.

Once the flask is successfully on board it is manoeuvred across the deck on the rail network to stop above the storage racks. The flask doors and the deck hatch doors are opened and the penetrator lowered to its position within the rack. Again hydraulic operated horizontal bolts are closed above the penetrators in the rack to form a seating for the one above.

The shielding is provided by thickening the sides of the ships hold and the deck structure accordingly.

Although this would be a slow loading process it has many attractive features, not least of which is that it can be engineered to high safety standards and that a very large proportion of the equipment requiring maintenance can be sited outside the shielded zones.

7.7.3 Ship and storage

The ship design for this Option given on Figure 40 is for 12 canister capacity penetrators. The design concept would be similar for reference penetrators and could permit simpler transverse storage.

The ship layout is conventional with many longitudinal full height bulkheads in the hold and transverse bulkheads at 20m. The moonpool is placed at the centre of the ship for stability.

7.7.4 Release

The two transfer flasks which would be required for speed and back-up, would remain onboard during the voyage. At release they perform the loading operation in reverse but this time they are positioned above the moonpool, the doors are opened and the penetrator is lowered through the moonpool so that the penetrator hangs beneath the ship. This would allow either horizontal or vertical launch attitudes to be selected.

The present concept does not allow the penetrator to be rotated in plan to point in any direction but this facility could be incorporated or the ship could be manoeuvred.

7.8 Option D

7.8.1 Summary

This option is based upon the design of a rail ferry recently commissioned. The penetrators are loaded through a shielded tunnel which mates with the stern of the ship. They are stowed aboard ship on carriages which are propelled by remotely controlled locomotives and distributed using lifts and traversers. All of the penetrator designs considered in this study could be handled by this option.

7.8.2 Loading

The penetrators are stored at the dockside in a shielded area, much like a marshalling yard for railway carriages. A gate in the yard building is designed to mate with the stern of a specially built ship whose stern door is lowered to form an articulated loading ramp. This system is commonly available for roll on roll off ferries. The penetrators are hauled on board using the ships locomotives and stowed by operating the ships lift and traversers.

7.8.3 Ship and storage

The configuration of the ship shown in Figure 41 is based on a tried and tested design. The locomotives are electro hydraulic and can be remotely controlled using closed circuit TV cameras for monitoring.

The central section of the stern is pivoted at the inboard end and can be raised and lowered by hydraulic arms at the outboard end. This ramp is activated to compensate for tidal and weight movements during loading and to provide a slipway during release which is discussed in the next section.

The absence of bulkheads in the holds is seen as a major disadvantage in this type of vessel as local damage could lead to uncontrolled flooding and possible capsize.

The loading ramp arrangement is located at the rear and requires the accommodation and engines to be at the bow.

7.8.4 Release

At the release station the locomotive propels a penetrator onto the rear ramp where it is clamped. The locomotive is withdrawn and the end of the ramp lowered until the outboard end is below sea level. The wheel clamps are released and the combination rolls into the sea. The nose carriage is shaped to restrain the penetrator on the inclined ramp but to detach itself in the water.

The carriages would not have brakes or springs but would have safety lines so that they could be recovered after the launch.

7.9 Option E

7.9.1 Summary

In this option shown in Figure 42, Penetrators are carried in transport flasks and stowed in conventional holds within the ship for the voyage to the disposal site. Each flask is then taken by a travelling portal crane to a launching frame. The launching frame upends the flask so that its penetrator can be released tail first through the ship's moonpool. The empty flask is replaced in the hold for the return voyage.

7.9.2 Loading

Assembled penetrators would be conveyed to the dock in transport flasks which would be loaded into conventional holds in the disposal ship by a dock crane. It is estimated that the total weight of such a flask and a reference penetrator would be between 100 and 120t and about 10m long. A flask of this size could readily be transported between an assembly plant and a disposal ship loading dock by train. This factor would be significant if only inland sites were available for assembly plants.

7.9.3 Ship and storage

The only novel aspects of the disposal ship would be the requirement for a high capacity travelling portal crane, a launching frame to handle the transport flasks and a moonpool.

The onboard handling equipment could probably be made to cater for a range of transport flask sizes and thus penetrator designs. Therefore, a disposal ship could accept penetrators of different sizes produced, say, by different countries. Such flexibility would be more expensive and difficult to achieve with the remote handling equipment within shielded holds which would be required for the other design options.

The radiation shielding of the transport flasks would be adequate for the relatively short periods that members of the crew would spend inspecting or attaching lifting tackle to them. However, it is likely that the ship would be built with a transverse flooded compartment or cofferdam between the flask holds and the crew accommodation areas.

7.9.4 Release

The penetrator release sequence would be:-

a) cover removed from one of the ship's holds by the travelling portal crane;

b) crane lifting hooks attached to trunions on a penetrator transport flask in the hold;

c) flask lifted out of the hold and transferred by the travelling crane to the moonpool launching frame;

d) flask, still in a horizontal attitude, secured on launching frame and crane moved away from the moonpool area;

e) flask lid transit bolts removed;

f) hydraulically powered launching frame operated to rotate the flask to the vertical. Then flask lowered into the moonpool until its mouth is adequately underwater;

g) flask lid hydraulically opened and internal clamps opened to release the penetrator when the ship is correctly positioned;

h) operations a) to f) undertaken in reverse sequence to replace empty flask in the ship's hold for the return voyage.

7.10 Conclusions

Having ruled out the idea of assembling penetrators on board ship it was necessary to establish the form of vessel which would be required to load up the assembled penetrators, take them out to the disposal site and successfully launch them. Other work conducted in this study has demonstrated that the angle of launch has an effect on the accuracy of emplacement.

The first task has been to demonstrate that the loading and launching idea was feasible and this has been done.

It soon became apparent that a number of options were reasonable and that the length and weight of the penetrator had a considerable effect.

In an attempt to cover the various options to ensure that no reasonable system was discounted, five systems have been developed and presented. Some of these appear to present considerable operational difficulties.

The length of the reference penetrator does not appear to present any particular difficulties apart from in Option B, which is not favoured for a number of reasons. Penetrators longer than the reference design, however, are difficult to accommodate across the ships beam which is a significant restriction.

At this stage Option E or a hybrid system derived from Options A and C would appear to have the most to recommend them.

8.0 DISPOSAL SITE USAGE

8.1 Introduction

This section presents information about seabed study areas and develops some of the data which would be required to produce preliminary plans for a penetrator disposal site.

8.2 Site Selection

As discussed in Section 1.4, the Site Selection Task Group of the NEA Seabed Working Group (SWG) have identified several deep ocean seabed areas as being suitable for further study in the context of radioactive waste disposal.

Amongst these, the area known as Great Meteor East (GME) in the Northeast Atlantic is one of the closest sites to Western Europe and has been adopted as the reference site for this study.

GME is situated between 30°N and 33°N, 23°W and 26°30W on the western margin of the Madeira Abyssal Plain. It has been studied in some detail since 1980 and site investigation material is reported in the annual proceedings of the SWG.

Generally, the site is an area of smooth horizontal seafloor with scattered abyssal hills. The average water depth is 5445m with the hills rising to approximately 500m above the flat-lying sediments of the plain. The overall character of the topography suggests partial burial of a hilly terrain by the sediments to the west and south, and an almost complete burial in the eastern part of the plain.

At this stage of the study, it would seem reasonable to choose emplacement areas away from the abyssal hills where the sediments have been found to be thickest. Part of the GME study area is shown on Figure 48 from which the locations of the abyssal hills which rise abruptly from the plain may be seen.

8.3 Factors Affecting Rate of Site Usage

The following two parameters have been identified as principally affecting the rate of site usage, assuming that reference penetrators are employed:-

(a) the rate of HLW disposal

(b) the penetrator drop spacing

8.4 Rate of HLW Disposal

The rate of disposal of HLW arisings will be taken as $85m^3$/yr which is equivalent to approximately 570 canisters/yr. Further information concerning the rate of disposal is given in Appendix A.

8.5 Penetrator Launch Spacing

The launch spacing for penetrators would be selected with the objective of achieving satisfactory spacing of emplaced waste. The problem of quantifying launch spacing therefore may be divided into two parts, namely

- design emplaced waste spacing
- launch spacing to achieve design emplacement spacing

8.5.1 Design emplacement spacing

To date most penetrator emplacement research has concentrated on embedment depth and little attention has been paid to spacing criteria.

At this stage it appears that spacing criteria would need to include consideration of low probability events, such as collision within the seabed, and more probable but possibly less serious situations such as thermal interaction.

These criteria are discussed further in Reference 21 which concludes that a design spacing of 15m would be appropriate for the reference penetrator used in this study. The methodology used to predict temperatures for emplaced groups of penetrators is given in Appendix G.

8.5.2 Design launch spacing

A number of independent factors would influence the achievement of the design embedment spacing. Those which have been considered in selecting the design launch spacing are as follows:

- the positional accuracy of the disposal ship
- the ships attitude (pitch, roll etc.)
- the penetrator launch attitude
- the penetrator manufacturing imperfections
- the ocean currents

These factors are considered individually in the following sections and their incorporation into a trajectory model is then discussed.

8.6 Ship Positioning Systems

The position of a ship can be given either precisely in earth co-ordinates or relative to another position, which is commonly on the coast line. In cases where the "known" position is on the seafloor, it is not usually necessary to know its exact earth co-ordinates. It is, therefore, necessary to decide how the disposal site area will be used before a positioning system can be chosen. Either an exact global positioning strategy can be adopted for each voyage or a relative positioning strategy taken up following an initial "marking" on the first voyage. Positioning systems are outlined in the following sections and are discussed in more detail in Reference 6.

8.6.1 Hyperbolic radio location systems

These systems can be high or low frequency and require special receivers to determine the phase difference between shore based master and slave transmitters. With a high frequency system such as "Loran-C" the range is limited to some 500-1000 km from shore, and this means that GME would be near the limit of its range. The position information is continuously available and the accuracy some 100-300m.

8.6.2 Long/short/super short baseline systems

All these systems work by measuring the time taken for a sound pulse to travel between a transducer on the ship and a transponder or group of transponders placed on the seafloor.

The accuracy is of the order of 1-2% of the water depth but decreases with the horizontal deviation of the bottom transponder. At 5000m depth this will give a vessel positioning accuracy of 50-100m at best, relative to the transponder.

8.6.3 Transit satellite system

This is a satellite navigation system based on a small number of orbital satellites. A special transit satellite receiver and data processor, connected to a special antenna are required. The equipment is currently widely used but readings are only obtained every 15-45 minutes, depending on the latitude. The claimed accuracy is 180-750m.

8.6.4 Navstar

Navstar is a military satellite navigation system which gives continuous positional information with a claimed accuracy of 10-30m. Higher accuracies are available for classified users only. It is expected to become available for civil applications in 1985.

8.6.5 Inertial navigation

Inertial navigation relies on accelerometers mounted on a 'gyro-stabilised platform'. The system is independent of any shore or space facility and relies purely on the integration of acceleration values to produce displacements. This continuous integration can lead to error accumulation and should be checked against known positions at regular intervals.

8.6.6 Scanning sonar systems

Modern sonar technology can be used as a navigational aid in exactly the same manner as surface radar. Long range sonars have ranges up to 20 km with resolutions of 50-100m.

8.6.7 Ship positional accuracy

The 10 - 30m positional accuracy claimed for Navstar has been taken as a reference value for this study.

8.7 Ship Attitude

The ship is unlikely to be exactly stationary during the launch of a penetrator. Any rolling, pitching or yawing motion of the ship would tend to give the penetrator an initial angle and angular velocity at launch which may cause an additional lateral displacement at the seafloor. Values of ±5° on angle and ±5°s^{-1} on angular velocity are assumed to be the limit of conditions under which the penetrators would be launched.

8.8 Method of Launch

The attitude and direction of the penetrator at launch will have a considerable effect on the lateral displacement at the seafloor. Vertical launching is considered to offer the greatest emplacement accuracy and is assumed to be used.

8.9 Ocean Currents

The maximum current speed at GME is assumed to be constant at 0.05 m/s below 2500m depth but to increase to 0.50 m/s at the surface.

Provided that the seafloor is relatively flat and stable any current should be effectively constant over a range of about 10 km.

Step currents might exist within 50 - 200m of the surface.

8.10 Manufacturing Imperfections

There are many areas where small imperfections in the manufacture of penetrators could occur including fin alignment, fin position and overall straightness. In addition, surface roughness and weight distribution are unlikely to be perfectly symmetrical. These imperfections could have a significant influence on penetrator trajectories and this is discussed below.

8.11 Penetrator Trajectories

When a free body falls through water the hydrodynamic and gravitational forces acting produce motion involving all six degrees of freedom. The motion in turn produces changes in the flow patterns around the body and a new set of hydrodynamic forces results.

The hydrodynamic forces produced by manufacturing imperfections, currents and launch attitude would lead to predictions of large lateral displacements if only planar motion were assumed. However, when all three dimensions are included in the assessment of motion, the trajectories are usually found to involve spiralling with only small overall displacements. Modelling studies (Ref. 21) have suggested that it would be advantageous to induce spinning by designed tail fin twists and that trajectories could be controlled such that lateral displacements of less than 15m could be effectively guaranteed for penetrators launched from nominally vertical attitudes.

These predictions have yet to be confirmed by trials data but have been adopted provisionally for this study.

8.12 Penetration Paths

There is some evidence to suggest that projectiles entering soils follow curved paths and in the case of penetrators this phenomenon could, for certain conditions, reduce the embedment depth achievable when compared with straight penetration.

For terrestial impacts Caudle (Ref. 20) states that the slenderness ratio of the projectile should be at least 8-10 to ensure stability. This would ensure that an incipient tumble could be resisted by the lateral afterbody forces which would be mobilised. He also suggests that the presence of fins on the tail is an aid to a stable underground trajectory.

Trajectories through seabed sediment have been modelled for a wide range of penetrators (Ref. 21) and deviations from the path at entry have been shown to be insignificant for most practical cases.

8.13 Rate of Site Usage

The nominal launch spacing used to assess the rate of site usage has been estimated as follows:

-	design emplaced waste spacing	7.5m
-	positional accuracy of disposal ship	30 m
-	penetrator trajectory in still water	15 m
-	ocean current allowance	10 m
		62.5m

Therefore the nominal launch spacing has been set at 125m. This assumes no knowledge of the locations of previously emplaced penetrators and ignores the fact that the current is likely to be constant and the allowance for it is therefore conservative.

The rate of seabed usage for 125m spacing is given below:

Volume of HLW per yr. (m^3)	No. of Reference Penetrators per yr.	Seabed Area Required per yr. (km^2)
0.75	1	0.016
85	115	1.9
170	230	3.6
400	540	8.5

The total extent of the GME study area is about 110,000 km^2. If only 10% is assumed suitable for waste disposal and the efficiency of site usage is only 10% of the values given in the Table above GME would still be adequate for over 500 years disposal at $85m^3$ per year.

8.14 Release Patterns for Voyages

The most efficient release procedure in terms of time and fuel costs would be to take straight line courses on each voyage and release the penetrators at intervals. The disposal site could then be simply divided into rectangular grids.

All options except B require the penetrators to be moved from storage to launch positions. Even the B option may require redistribution of ballast as the penetrators are released and for all the options, it is likely that some monitoring of the penetrators will be required to ensure successful emplacement.

For these reasons the ship would have to hold station over the required drop positions and for improved stability and manoeuvrability it would be preferable that the ship should head into wind. However, disposal operations could take place throughout the year and preliminary analysis of ocean wave statistics suggests that there will be a seasonal variation in the direction. It would, therefore, be uneconomic to allow sailing into the wind on each voyage regardless of the previous direction followed, and it is suggested that the grids be aligned with the prevailing wind direction. The effect of currents can be countered by the thrusters on the ship.

If a penetrator should fall outside the expected trajectory limits then the subsequent release pattern could be altered so that the final distribution of emplaced penetrators was acceptable. An operational plan would be prepared indicating what action should be taken in the event that any penetrator trajectory fell outside of preset limits. The plan would be developed on the basis of radiological assessments of non standard events.

8.15 Penetrator Monitoring

After each penetrator had been released it would be necessary to confirm its emplaced location, embedment depth and possibly recover other data concerning its in situ condition.

The information required could be gathered by the disposal vessel using instrumentation mounted on each penetrator and survey techniques. Longer term data could be gathered by survey vessels or ships on subsequent disposal cruises.

The exact requirements and methodology for monitoring cannot be specified at present but there appears to be no technical reason to doubt the feasibility of monitoring at least the location and depth of emplaced penetrators.

8.16 Emplacement and Transport Accidents

The method of emplacing HLW using free-fall penetrators is expected to be highly reliable because of its simplicity. However, several scenarios for unsuccessful emplacement have been assessed to evaluate the consequences. These primarily cover operational accidents which could occur during the transport of HLW to a seabed disposal site and during emplacement of the waste in the sediments.

8.16.1 Emplacement accidents

If one or all of the fins were to break off for some reason, then the trajectory would become highly unstable. However, in order to determine the minimum impact velocity, the penetrator has been assumed to fall in a stable horizontal attitude. The resultant velocity for a reference penetrator would be 6m/s and further simple calculations estimate the sediment penetration to be approximately 1.0m. It should be noted that the strength of the surface layers is very difficult to predict and hence the penetration figure is indicative only. However, burial appears to be a distinct possibility.

The probability of hitting a glacial erratic or any other solid obstruction is extremely low. The probability of a surface impact between a 0.98m diameter penetrator and a 0.06m diameter erratic, which is the largest recovered from the surface of soft ocean sediments, has been estimated to be 4.1 x 10^{-4} (Ref. 7). The probability of the same size penetrator impacting a 0.2m diameter erratic embedded 0.7m below the surface was estimated to be 3.3 x 10^{-5}. These probabilities should be regarded as "order of magnitude" figures in view of the limited data but in neither of the examples given would the embedment depth have been expected to be significantly reduced.

8.16.2 Transport accidents

The nature of the cargo and hence the safety features incorporated in handling and storing the penetrators makes the probability of accidentally launching penetrators in coastal waters very low.

The chances of sinking are considered to be exceedingly small particularly since nuclear waste disposal vessels would be routed well away from any busy shipping lanes.

For this reason only sinking in relatively shallow water is considered.

8.17 Penetrator Retrieval

There are two primary cases to consider:

(a) retrieval of penetrators from deep water due to unsuccessful emplacement;

(b) retrieval of penetrators from relatively shallow water following sinking of the disposal vessel or accidental launching.

It is necessary to assess the practicability of retrieval and the radiological consequences of non-retrieval in both cases.

8.17.1 Retrieval following unsuccessful emplacement

It has been estimated that partial burial of a penetrator is probable no matter how it falls to the seafloor. Although detection of the penetrator should not be too difficult since it may well be instrumented, it would appear extremely difficult to recover.

It is therefore considered unrealistic with current technology to attempt to recover a penetrator from unsuccessful emplacement. The situation would need to be reviewed should future studies show that there are in fact serious radiological implications arising from such an event.

8.17.2 Retrieval from shallow water

Shallow water detection is not anticipated to be a major problem and the attachment of recovery cables should be possible using submersibles provided there is no significant embedment. The capability of retrieving the whole ship either by lifting or re-floating is also considered feasible.

There would be a considerable amount of time available to mount a retrieval operation because the penetrator body provides at least 500 years of total containment. This assumes that there is no significant damage to the penetrators as a result of the accident which caused the ship to sink which seems reasonable for the class of ship proposed.

8.18 Conclusions

The NEA Seabed Working Group has identified several sites in the North Atlantic as being potentially suitable for radioactive waste disposal. One of these, Great Meteor East, has been chosen as a reference site for this study and has been shown to be sufficiently large for at least 500 years usage for reference disposal operation conditions.

From the results of trajectory modelling studies and ship positioning accuracy and attitude considerations, a nominal spacing of 125m between penetrator launch points has been proposed. The need to confirm this spacing by instrumented trials is acknowledged.

Aspects of penetrator monitoring and operational accidents have been discussed.

There appears to be no reason to doubt the feasibility of the penetrator option on the basis of site usage considerations although a number of areas would require development before feasibility could be finally confirmed.

9.0 COST ASSESSMENT

9.1 Introduction

The economic aspects of ocean disposal of HLW using the penetrator option are of less current importance than the engineering feasibility. However, it is important to give approximate indications of cost to ensure that these do not form an unreasonably large part of the corresponding electricity generating costs.

The aim has been to gauge the order of magnitude of the disposal costs and report these as the cost per canister disposed and also as the cost per unit of electricity produced.

Another important aspect is to determine the relative significance of the constituent costs. This would, for example, allow operating and capital costs to be compared and the importance of raw material costs to be assessed.

Two disposal systems have been costed to allow the effects of the assumptions made concerning penetrator design and operation to be tested.

The Reference Option is a combination of all the reference designs and reference operations discussed during the study.

The Conservative Option includes designs and operations based on more cautious assumptions than were used for the corresponding reference designs and operations.

Finally the approximate nature of the cost information cannot be overemphasised. It has not been used to optimise the designs nor to help select a preferred design.

9.2 Reference Option

The Reference Option has been developed to suit the following specification for HLW disposal.

a) disposal of $85m^3$ vitrified HLW annually.

b) HLW in canisters stored, nominally for 50 years, in an air cooled facility adjacent to the vitrification plant until ready for disposal. Approximately 570 canisters would have to be handled annually.

c) reference design of penetrator used for disposal. Nominally 115 penetrators would be required annually.

d) Great Meteor East seabed study area is assumed to be the disposal site for the reference cost operation.

9.2.1 HLW canister transport flasks

The procedure described in 6.9.2 for the transport of HLW canisters from the vitrification plant store to a penetrator assembly plant has also been assumed for the reference cost operation. The main aspects of the canister supply sequence can be outlined as follows.

a) HLW canisters taken by rail from the vitrification plant store to a penetrator assembly plant in transport flasks weighing about 120t. A flask would hold about 20 canisters so, nominally, 29 flask movements would be necessary per year.

b) it is considered that six train loads per year would prove most appropriate with regard to the efficient utilization of loading and unloading facilities. A nominal 100km distance by rail between the vitrification plant and assembly plant sites has been assumed. Thus a minimum of five flasks would be required. However, a total of six transport flasks and associated railway wagons has been assumed for the cost assessment to allow for maintenance and other contingencies. It has been assumed that flask maintenance facilities would be incorporated on the assembly plant site.

9.2.2 Penetrator assembly plant

The assembly plant described in Section 6.8 et seq and shown in Figures 28 to 33 has been assumed for the reference cost assessment. To summarise, two fundamental functions would be performed at the assembly plant as follows:-

a) canister preparation entailing ullage filling

b) penetrator assembly entailing canister loading, void filling and welding to complete penetrators.

The equipment and procedures for any necessary canister or penetrator rectification, plant maintenance and other services have also been considered in the cost assessment.

Auxiliary items such as boundary fencing, miscellaneous outbuildings, railway spurs, access roads, effluent treatment works, power supply and other infrastructure installations, and a facility for maintaining transport flasks are not shown on the assembly plant figures but have been allowed for in the cost assessment exercise.

9.2.3 Disposal ship dock facility

It has been assumed that dedicated docking facilities would be required for loading penetrators onto disposal ships. These facilities could be either specially constructed on a previously unused site or provided by redeveloping a more appropriately located existing harbour.

9.2.4 Disposal ships and penetrator transport flasks

For the cost assessment, the assembly plant has been assumed to be located at the dockside as this enables any penetrator size or ship option to be included.

The ship option C as shown in Figure 40 will be used for the reference cost calculation but will be assumed to carry 84 reference penetrators rather than the 42 longer penetrators shown in the Figure.

Self-propelled transfer flasks for loading and launching have been allowed for as described in Section 7.7.

9.2.5 Disposal site usage

Options for the management and use of seabed disposal sites were discussed in Section 8.0. No particular specification for disposal site management procedures has been set out for the reference cost operation because the options are insufficiently defined for meaningful costs to be assigned.

9.3 Conservative Option

The Conservative Option can be summarised as the disposal of HLW in two canister capacity, 200mm steel wall thickness penetrators. The following notes outline the upper bound operation in more detail:-

a) $85m^3$ per year HLW disposal rate (same as reference operation).

b) HLW canister operational details same as for reference operation.

c) Nominally, 285 penetrators would be required annually. This compares with the reference operation which assumes 115 reference penetrators per year. The two canister capacity penetrators would weigh about 22t because of their much greater steel wall thickness.

d) Southern Nares Abyssal Plain (SNAP) seabed study area as shown on Figure 1 assumed as disposal site.

e) HLW canisters supplied to the assembly plant in the same transport flasks as would be used for the Reference Option.

f) one assembly plant of the same cost as that considered for the reference operation plant has been assumed.

g) one option B design of ship and dock facility as shown in Figure 39 and described in Section 7.6 has been assumed. Three voyages per year would be necessary.

h) it has been assumed that the assembly plant would be built alongside the disposal ship loading berth so that only five flasks would be needed to transfer penetrators between the plant and the ship.

9.4 Costing Principles

The capital, operating and omitted costs considered for the assessment exercise and the method use for combining them are explained in the following section.

9.4.1 Capital costs

The capital costs considered in the assessments have been divided between two amortization periods. A discounting period of 25 years has been taken for relatively durable installations such as buildings. Items with a significant mechanical component, disposal ships for example, have been assumed to have a shorter effective service life of 12½ years over which their initial cost must be paid off. Maintenance and other recurring charges associated with capital cost items have been allowed for in the compilation of annual operating costs as outlined in 9.4.2.

The items included as capital costs at March 1984 price rates are as follows:-

a) 25 year amortization:

- Assembly plant buildings including infrastructure works
- Disposal ship dock facility wharfs and buildings including infrastructure works
- HLW canister transport flasks
- Penetrator transfer flasks

b) 12½ year amortization:

- Assembly plant mechanical equipment
- Assembly plant electronic equipment
- Assembly plant fittings
- Disposal ship dock facility mechanical equipment electronic equipment and fittings
- Disposal ships.
- Railway wagons for transport flasks.

9.4.2 Operating costs

The annual operating costs of the following items at March 1984 price rates have been included:

Assembly plants:

- Transport of HLW canisters to the assembly plant.
- Fabricated steel penetrator components.
- Supplies for penetrator completion.
- All other supplies for the routine operation and maintenance of assembly plants.
- Salaries of assembly plant employees including benefits, insurance and pension commitments.

Dock facilities:

- Supplies for the routine operation and maintenance of dock facilities.
- Salaries of dock facility employees including benefits, insurance and pension commitments.

Disposal ships:

- Supplies for the routine operation and maintenance of disposal ships.
- Salaries of crew members including benefits, insurance and pension commitments.

9.4.3 Omitted costs

At this early feasibility study stage it is not possible to specify all the facets of the HLW penetrator disposal programme in the detail required for a full cost assessment to be made. For example, several aspects are, as yet, so ill defined that meaningful estimates of their costs cannot be made. Also, it is not appropriate for certain other aspects to be included in cost assessments at this time. Some of the items omitted from the cost estimates are as follows:-

- Initial development work such as feasibility studies, research programmes, prototype testing and inquiries.
- Continuing development work undertaken after the commissioning of a penetrator disposal programme to identify operational improvements.

- Interim storage of HLW canisters and special facilities for loading HLW canisters into flasks for transport to penetrator assembly plants.

- Purchase of land for assembly plants and disposal ship dock facilities.

- Servicing of any loans needed to cover stage payments for initial development and construction work.

- Seabed disposal site establishment and monitoring and any instruments which may be required on penetrators.

- Charges levied by any national or international organisation that may be appointed as independent assessors or supervisors for a disposal programme.

- Recovery ships and equipment that may be specified as a precaution to be available to retrieve any penetrators accidently lost at sea.

- Decommissioning of assembly plants, dock facilities and disposal ships.

9.4.4 Combined costs

In order to express total costs on an annual basis a nominal figure of 10% has been assumed as the continuously compounded real rate of interest that would be required on loans arranged for initial investments in capital cost facilities and equipment.

9.5 Cost Estimates

The estimated costs of the Reference and Conservative Options outlined in 9.2 and 9.3 are given in this section. Costs are summarised in Table 9.1 and a breakdown is given in Table 9.2.

It is stressed that it is far too early to select options for HLW disposal on grounds of cost. The estimates, therefore, only give a very general comparison of the cost implications of alternative schemes.

9.5.1 Reference option

a) £129m estimated as capital cost of:-

- six canister transport flasks and railway wagons
- one assembly plant
- one dock facility
- one disposal ship option C.

b) £11m estimated as annual operating cost of:-

- six train journeys to take canisters in transport flasks to the assembly plant
- steel components for 115 reference design penetrators
- supplies for completing penetrators
- other supplies to the assembly plant and dock facility and for two disposal voyages
- staff costs

c) £27m calculated as combined capital and operating cost on an annual basis.

9.5.2 Conservative option

a) £135m estimated as capital cost of:-

- six canister transport flasks and railway wagons
- one assembly plant
- one dock facility
- five penetrator transfer flasks
- one disposal ship option B

b) £22m estimated as annual operating cost of:-

- six train journeys to take canisters in transport flasks to the assembly plant
- steel components for 285 two canister capacity 200mm wall thickness penetrators
- supplies for completing penetrators
- other supplies to the assembly plant and dock facility and for three disposal voyages
- staff costs

c) £39m calculated as combined capital and operating cost on an annual basis.

9.6 Cost Sensitivity

The sensitivity of the estimated costs to variations in the principal assumptions made in the Reference Option is discussed in the following sections.

9.6.1 HLW disposal rate

There is no definitely planned rate for HLW disposal so three representative cases have been assumed. They are:-

a) $85m^3$ HLW/yr
b) $170m^3$ HLW/yr
c) $400m^3$ HLW/yr

In general, the unit cost of handling HLW canisters would be less for the higher rates of disposal because of the greater dilution of overheads such as initial research and construction and subsequent administration expenses. However, the implication is that higher disposal rates would involve handling HLW from several relatively widely dispersed vitrification plants. Therefore, there would be increased costs due either to the transport of HLW canisters over long distances to a single major penetrator assembly plant or the construction of several small assembly facilities adjacent to each vitrification plant.

It is thus concluded that HLW penetrator disposal schemes cannot yet be defined in the detail necessary to discern meaningful variations in the unit costs of operations for different rates of disposal.

9.6.2 HLW age at disposal

The method adopted for cost assessments does not consider the influence of HLW age at disposal. It is suggested that any chosen option for disposal would handle consignments of HLW at as low an age as feasible to minimise storage facilities and thus costs at vitrification plants. It is likely to be some time before sufficient information is available to identify a minimum appropriate age for HLW at disposal. Consequently a detailed analysis of cost sensitivity on this matter cannot yet be undertaken. However, it is not felt that there would be unduly large cost implications in comparison with the total cost of a disposal programme.

9.6.3 HLW canister design

Only the Figure 2 canister design has been considered in the cost assessments. Slightly greater disposal costs would be incurred if smaller canisters were chosen.

9.6.4 Penetrator design and manufacture

At this stage, cost assessments indicate that penetrator components and materials such as void fillers could account for about 50% of the total costs of a disposal operation. Optimisation of penetrator design, therefore, would be central to the development of a cost effective HLW disposal scheme.

Once a choice has been made between, say, the adoption of corrosion resistant or allowance metals, costs are effectively proportional to the amount of material used for penetrator construction. Therefore, studies should concentrate on identifying the minimum penetrator weight per canister that would be appropriate. Wall thickness selection and the design of nose and tail pieces are most important in this respect. Canister capacity, that is the length, of penetrators is also significant because the number of nose and tail pieces needed for a given disposal programme is reduced if greater canister capacity penetrators are chosen.

9.6.5 Assembly plant, dock facility and disposal ship design

Assembly plants, dock facilities and ships would be the principal capital cost items of a penetrator disposal operation. At this stage it is only possible to make very approximate costings based on conceptual designs. The specification and pricing of penetrator assembly equipment is especially uncertain so far in advance of the development of firm technical requirements for penetrator design and construction. Clearly the cost assessment of a disposal operation is particularly sensitive to assumptions made about the major capital cost items. However, it will not be feasible to consider this in detail until it is possible to more closely define the form that HLW disposal schemes would take.

9.7 Conclusions

It is useful for likely orders of costs to be expressed in terms of the disposal cost per HLW canister or some other common factor when comparing the penetrator option with alternative proposals such as land repositories. The conversions given in Appendix B have been used to prepare the cost information in this manner as presented in Table 9.1.

It is noted the estimated HLW disposal cost per kW(e)hr for the reference operation of 0.007p compares with the current domestic charge rate in the UK of around 5.0p.

10.0 FEASIBILITY ASSESSMENT AND RECOMMENDATIONS

10.1 Introduction

A number of different systems for disposal of HLW using penetrators were considered in this study and the major features of each are set out in Part 1 of this report. The most promising options were selected for study in greater detail and that work is described in Part 2 of this report.

This feasibility assessment is based on the arguments and information presented in Part 2 of the report.

10.2 Reference Designs

Reference designs and reference sequences of operations have been proposed for penetrators, disposal ships and penetrator assembly plants. They have not been sufficiently developed to be considered candidate designs should this method of disposal ever be adopted but they contain adequate detail of the important features to allow the feasibility of the penetratror disposal concept to be assessed.

In particular, the penetrator wall thickness has been selected following consideration of corrosion data generated from short term experimental programmes. Further research is required on corrosion prediction and acceleration mechanisms such as embrittlement and cracking need particular attention.

10.3 Pentrator Component Fabrication

There is no doubt that the components required to build penetrators could be made by a number of present day processes to the specified quality. Component production would involve primarily readily available low carbon steels and could be carried out by manufacturers outside the nuclear industry as there would be no radiation hazard.

10.4 Penetrator Assembly

The essential operations involved in assembling a penetrator would be to weld together the nose, body-tube and tail components and to completely fill the internal spaces with lead, or other suitable material.

These operations must satisfy stringent quality standards to ensure a very high integrity product and have to be performed remotely in a heavily shielded assembly cell.

The preferred welding techniques are often carried out automatically in general fabrication shops and some are also performed in active areas of nuclear process plant. Examples of high integrity filling operations in a radioactive environment are less easy to find but there appears to be no reason why they should not be feasible.

Penetrator assembly is therefore considered technically feasible but requires operational development before complete feasibility can be confirmed.

10.5 Disposal Operations

Ships are currently used for the international safe transport of irradiated nuclear fuel, hazardous chemicals and munitions and little further development of the safety features incorporated in these vessels would be required for the waste disposal operations considered in this study. However handling of penetrators prior to launching would need demonstration before its feasibility could be confirmed.

Loading the penetrators onto the disposal vessel at a dedicated dock is, by comparison with launching operation relatively straightforward.

There appear to be no major logistical problems with the type and scale of the operations envisaged.

10.6 Site Usage and Monitoring

The extent and characteristics of some of the North Atlantic seabed study areas appear adequate for European HLW disposal requirements for many years.

A nominal launch spacing of 125m between penetrators designed to meet certain reasonable criteria is predicted to be enough to limit the probability of thermal interaction or collisions to an insignificant level.

The calculated peak in situ temperature for the sediment adjacent to a reference penetrator carrying five canisters of HLW which have been stored for fifty years before disposal is about 170°C. At present sediment response to elevated temperatures is not well understood and the possibility of loss of integrity of the sediment barrier cannot be ruled out. Further research is needed in this area and the alternative courses of longer waste storage or lower waste concentration in the penetrator should be considered. The peak temperature calculated for the equivalent case but with 100 years stored waste is about 60°C.

10.7 Feasibility Conclusions

There appears to be no single technical or operational reason when judged by reasonable engineering criteria, to doubt that the penetrator option for HLW disposal would be feasible. However, to increase confidence in this conclusion further research, study and demonstrations should be carried out into corrosion, active assembly operations, penetrator handling at sea, emplacement trajectories and the thermal response of sediments.

REFERENCES

1. Department of The Environment (1977) — "Nuclear Power and the Environment" The Government's Response to the Sixth Report of the Royal Commission on Environmental Pollution (Cmnd. 6820) HMSO.

2. Valent P.J. and Lee H.J. (1976) — Feasibility of Subseafloor Emplacement of Nuclear Waste, Marine Geotechnology, Vol. 1, No. 4 pp. 267-293.

3. Silva A.J. (1977) — Physical Processes in Deep-Sea Clays, Oceanus Vol. 20, No. 1, pp 31-40.

4. Atkins Planning (1981) — Concepts for the Disposal of High Level Radio-active Waste: Task 7. The Deep Ocean Bed, DoE Report No. DoE/RW/82.015.

5. Kenter C.J. and Visser P.J. (1981) — A Tentative Study of Soil Mechanics Aspects of Emplacement of Radioactive Waste Canisters in Deep Sea Sediment Layers, Delft Soil Mechanics Laboratory, Report No. BO-258680/14.

6. Manschot D. and Stein Callenfels, J.E. van, (1981) — A Survey of Emplacement Methods for Radioactive Canisters in Deep Sea Ocean Sediments. Marine Structure Consultants report to DORA, Holland.

7. Ove Arup & Partners (1982) — Ocean Disposal of High Level Radioactive Waste - Penetrator Option. Studies Relevant to Emplacement in Deep Ocean Sediments. DoE Report No. DoE/RW/82.102.

8. Dayal U. and Allen J.H. (1975) — The Effect of Penetration Rate on the Strength of Remoulded Clay and Sand Samples, Canadian Geotechnical Journal, vol. 12, pp 336-348.

9. Beard R.M. (1977) — Expendable Doppler Penetrometer: A Performance Evaluation. Civil Engineering Laboratory, Technical Note R-855, Port Hueneme, California.

10. Ove Arup & Partners (1983) — Ocean Disposal of High Level Radioactive Waste, Penetrator Engineering Study Phase 1, Preliminary Feasibility. DoE Report No. DoE/RW/83.094.

11. Freeman T.J. et al (1984) — Modelling Radioactive Waste Disposal by Penetrator Experiments in the Abyssal Atlantic Ocean, Nature, Vol. 310, (July) pp 130 - 133.

12. OECD Nuclear Energy Agency (1984) — Seabed Disposal of High-Level Radioactive Waste. A Status Report on the NEA Coordinated Research Programme.

13.	HMSO (1982)	Radioactive Waste Management: Department of the Environment: July 1982: Command 8607
14.	Entwicklungsgemeinschaft Tieflagerung (1984)	Trial Storage of High-Level Waste in the Asse II Salt Mine. Commission of the European Communities. Nuclear Science and Technology, EUR 8796 EN.
15.	Hill M. D. and Delow C.E. (1982)	Interim Meeting of the SWG Systems Analysis Task Group, Baden, October 1982, in SAND 83-2122.
16.	Rothwell G.P. (1981)	Corrosion Problems Related to the Containment of High-level Nuclear Waste for Disposal, DoE Report No. DoE/RW/81.002.
17.	Savvidou C. (1984)	Effects of a Heat Source in Saturated Clay, Ph.D. thesis, University of Cambridge.
18.	Anderson D.R. (1984)	Proceedings of the 8th Annual Meeting of SWG, Varese, 1983, Sandia National Laboratories, SAND 83-2122.
19.	International Commission on Radiological Protection (1977)	Annals of the ICRP, Vol. 1, No. 3, ICRP Publication 26 (Radiation Protection), Pergamon Press.
20.	Caudle W.N. et al (1967)	The Feasibility of Rapid Soil Investigations Using High-speed Earth-Penetrating Projectiles. International Symposium on Wave Propagation and Dynamic Properties, ASCE, Albuquerque, pp 945 - 955.
21.	Ove Arup and Partners (1985)	Ocean Disposal of Heat-Generating Radioactive Waste, Penetrator Trajectory Modelling and Emplacement Criteria, DoE Report, to be published.
22.	Ove Arup and Partners (March 1982)	An Initial Study into Penetrator Shape and Size for Emplacement in Deep Ocean Sediments, DoE Report No. DoE/RW/82.055.
23.	Canfield T. and Burchett S.N. (1982)	Hole Closure Calculations Using HONDO II and MARBLE. Subseabed Disposal Program, Annual Report Jan-Sept 1981. SAND82-0664/1, Volume II, Part 2, Appendix FF, pp 301-334.
24.	Dzwilewski P.T. (1982)	Hole Closure Behind a Deep Ocean Sediment Penetrator. Subseabed Disposal Program, Annual Report, Jan-Sept 1981 SAND82-0664/1 Volume II, Part 2, Appendix GG, pp 335-358.

25. Marsh G.P. (1982) — Materials for HLW Containment, Nuclear Energy, Vol 21, No 4, pp 253-265.

26. Accary A. (1984) — Corrosion Behaviour of Container Materials for Geological Disposal of High Level Waste, Commission of the European Communities, Nuclear Science and Technology, EUR 9570.

27. Nuttall A. and Urbanic V.F. (1981) — An Assessment of Materials for Nuclear Fuel Immobilisation Containers. AECL-6440.

28. Westinghouse Electric Corporation (1982) — Engineered Waste Package Conceptual Design Defense High-Level Waste (Form 1), Commercial High-Level Waste (Form 1) and Spent Fuel Disposal in Salt, Office of Nuclear Waste Isolation, ONWI 438.

29. Teper, B. (1980) — Particulate Compaction Tests for a Particulate Packed Thin-wall Container for Irradiated Fuel Disposal, Atomic Energy of Canada Limited, AECL TR131.

30. Rosinger E.L.J. and Dixon R.S. (1982) — Fourth Annual Report of the Canadian Nuclear Fuel Waste Management Programme, Atomic Energy of Canada Limited, Whiteshell.

31. Canadian Nuclear Society (1982) — International Conference on Radioactive Waste Management Winnipeg, 12-15th Sept. 1982.

32. International Atomic Energy Authority (1979) — Regulations for the Safe Transport of Radioactive Materials. 1973 Revised Edition (As Amended). Safety Series No. 6, IAEA, Vienna.

33. Hills P.J. (1982) — Relative Dangers of Carrying Fuels in Britain by Inland Surface Modes: The Risks of Fuel Transport Conference Proceedings. Hazardous Cargo Bulletin.

34. Burton W.R. and Griffin J.R. (June 1981) — A Design Study of Long Term Storage and Underground Disposal Systems for Highly Active Waste. UKAEA Northern Division Report ND-R-514 (R).

35. Kempe T.F. and Martin A. (May 1981) — Pre-disposal Storage, Transport and Handling of Vitrified HLW. DoE Report No. DoE/RW/82.016.

36. British Standards Institute (1967) — Dimensional and Quantity Tolerances for Steel Drop and Press Forgings and for Upset Forgings Made on Horizontal Forging Machines, BS 4114.

37. British Standards Institute (1980) — Specification for a Total Quality Assurance Programme for Nuclear Power Plants, BS 5882.

38. International Atomic Energy Authority (1978) — Quality Assurance for Safety in Nuclear Power Plants, Code of Practice.

39. HM Nuclear Installations Inspectorate — Guide to the Quality Assurance Programme for Nuclear Power Plants.

40. British Standards Institute (1979) — Quality Systems, BS5750.

41. Steel Castings Research and Trade Association (1959) — A Practical Guide to the Design of Steel Castings

42. Taylor, K.J. et al (1983) — Corrosion Studies on Containment Materials for Vitrified High Level Nuclear Waste, DoE Report No. DoE/RW/83.058.

43. Leggatt R.H. (1982) — Residual Stresses at Circumferential Welds in Pipes, Welding Institute Research Bulletin.

44. Nichols R.W. ed. — Developments in Pressure Vessels (1979) Technology - 1 (Flow Analysis), Applied Science Publishers Ltd.

45. Nichols R.W. (1983) — Ultrasonic NDT Techniques, Nuclear Energy, Vol. 22, No. 4.

46. British Nuclear Energy Society (1983) — Defect Detection and Sizing, Nuclear Energy, Vol. 22, No. 5.

47. Hogben N. and Lumb F.E. (1967) — Ocean Wave Statistics, HMSO.

48. U.S. Navy (1955) — US Navy Marine Climate Atlas of the World Vol. 1, N. Atlantic.

49. Granville D.S. (June 1976) — Elements of the Drag of Underwater Bodies, Published by the David W. Taylor, Naval Ship Research and Development Centre, Ship Performance Department.

50. Hoerner (1965) — Fluid Dynamic Drag, Published by the author.

51. Hodgkinson D.P. and Bourke P.J. (1978) — The Far Field Heating Effects of a Radioactive Waste Repository in Hard Rock. Proceedings of the Seminar on In-situ Heating Experiments in Geological Formations, OECD/NEA, Ludvika/Stripa 1978.

52. Anderson D.R. (ed) (1981) — Proceedings of the Sixth Annual NEA-Seabed Working Group Meeting, Paris, France, February 2-5, 1981, p 187, Sandia National Laboratories, SAND81-0427.

Table 2.1

Corrosion Behaviour Classification of Materials

A: Cast Irons and Carbon Steels

Material	Description (1)	Corrosion Rates (2)	Remarks
Cast irons	1.7 to 6.7% carbon	High to Medium (for low oxygen supply) general and local corr.	White and Grey C.I. brittle but malleable (long heat treatments) and spheroidal graphite (cerium or magnesium additions) C.I. ductile. Unsuitable for welding. Widely available.
Plain carbon and C-manganese steels	0.1 to 1.7% C. up to 2% Mn. up to 0.5% silicon	High to Medium (for low oxygen supply) general and local corr.	Good mechanical properties Low C (<0.25%) steels weldable, Medium C (<0.45%) and High C steels increasingly difficult or unsuitable for welding. Widely available.
Very low carbon steels	less than 0.1%C. up to 2% Mn. up to 0.5% Si.	High to Medium (for low oxygen supply) general and local corr.	Under development for engineering applications so availability and data limited. Similar to low C steels but may be easier to weld.
Pure iron (Fe)	less than 0.03%C.	High to Medium (for low oxygen supply) general and local corr.	No engineering applications so availability and data very limited.

Table 2.1 (cont'd)

B: Stainless Steels

Material	Description (1)	Corrosion Rates (2)	Remarks
Ferritic stainless steels(FSS)	15 to 30% Chromium	Low general and local corr. in appropriate environment and services conditions.	Poor mechanical properties at about room temperature. Used in high temperature corrosive environments. Low susceptibility to SCC (3). Very prone to embrittlement and local corrosion at welds.
Austenitic stainless steels (ASS)	more than 8% Nickel, more than 8% Cr, and at least 24% (Ni + Cr).	As above.	Better mechanical properties and corrosion resistance so more applications than FSS. Susceptible to SCC. Stabilized ASS (Titanium or Niobium additions) generally weldable.
Martensitic stainless steels (MSS)	12 to 18% Cr. up to 3% Ni. 0.1 to 1%C.	As above.	Heat treatable for improved strength but less corrosion resistance than other SS. Susceptible to SCC Prone to weld embrittlement.

Table 2.1 (cont'd)

C: Copper, Titanium and high Nickel Alloys

Material	Description (1)	Corrosion Rates (2)	Remarks
Copper and Cu alloys	Alloys incl. brasses, bronzes, Al, Ni, Mn and Si.	Low general and local corr. within environment range in which stable passive film develops.	Low yield strength and creep resistance Weldability varies from poor for brasses to reasonable for copper and Cu-Ni alloys.
Titanium and alpha Ti-alloys	Alloys incl. Al, Ni, Pd and Sn.	as above.	High strength to weight ratio. Low high temperature strength. Low Hydrogen content necessary to avoid embrittlement. Weldable with precautions
Incoloy	Fe-base 30 to 40% Ni 21% Cr.	as above	Developed for very high temperature corrosion resistance Weldable with precautions.
Monel	Ni-base 25 to 30% Cu	as above	Developed for good mechanical properties and corr. resistance at high temperatures. Weldable with precautions.
Inconel	Ni-base. 16% Cr. 8% Fe.	as above	Generally superior properties than Monel. Weldable with precautions.
Hastelloy proprietary Ni-Alloy range	Alloys incl. Cr, Cu, Fe, Mo, Si and W.	as above	Developed for good mechanical properties and corr. resistance at high temps. for specific applications. Weldable with precautions

Table 2.1 (cont'd)

D: Miscellaneous Materials

Material	Description (1)	Corrosion Rates (2)	Remarks
Alloy steels	Carbon steel based alloys	Generally similar to plain C steels	Low and high alloy steels developed for particular engineering requirements (e.g. high strength, low temperature ductility, high machineability, low creep).
Precipitation hardening duplex stainless steels	Fe-based Cr.Ni complex alloys	Generally similar to austenitic stainless steels (ASS)	Developed for good mechanical and properties and corr. resistance at high temperatures. Some grades less susceptible to SCC than ASS. (3).
Aluminium and Al-alloys	Alloys incl. Cu, Mg, Mn, Si, and Zn	High general and local corr. outside the limited environment range in which stable passive film develops.	Developed for engineering applications where high strength to weight ratio required.
Lead	100% Pb	Generally low but can be high in warm seawater	Very low strength and creep resistance

Notes:

(1) Descriptions exclude impurities and small alloy additions

(2) Medium corrosion rate implies a material thickness loss during 500 years seabed embedment of the order of 50mm. Low and high corrosion rates imply corresponding material thickness losses of significantly less and significantly more than 50mm. Predictions fluctuate widely depending on material quality, disposal environment, etc. so information in Table must be considered as conjectural pending the results of future site and operation specific research.

(3) SCC is abbreviation of stress corrosion cracking. Only the relative resistance to SCC of the various stainless steels is given although most of the other materials are also susceptible to SCC in adverse conditions.

Table 2.2

Densities and Yield Strengths of Materials

Material	Typical Density (kg/m³)	Yield Strength (N/mm²)	
		Nominal Range	Typical(1)
Cast Irons	7700	200 - 500	200
Carbon steels	7850	200 - 400	200
Stainless St.	7950	200 - 1000	200
Cu-alloy	8900	20 - 200	100
Ti-alloy	4550	200 - 800	300
Incoloy	8200	200 - 600	200
Monel	8850	150 - 600	200
Incoloy	8500	200 - 700	250
Hastelloy (2)	8950	-	300
Alloy steels	7850	200-1000	400
Al-alloy	2650	20-350	100
Lead	11350	-	-

Notes:

(1) Yield strengths fluctuate widely depending on specified alloy composition and condition of material, etc. so data in Table must be treated as only approximate guides to likely room temperature values. Yield strengths of some materials, e.g. steels and titanium, can have pronounced reduction at around the typical peak temperature limit of 100°C for penetrators following seabed embedment.

(2) For Hastelloy C grades of alloy range.

Table 2.3

Relative Costs of Materials

Material	Relative Cost Guide (Steel = 100)		Relative Thickness of plates
	per tonne	per cu.metre	of Equal Cost (Ti-alloy = 5)
C.I. or C-Steel	100	100	49
Stainless St. (3)	500	505	10
Cu-alloy	500	570	9
Ti-alloy	1700	985	5
Incoloy (3)	800	835	6
Monel (3)	900	1015	5
Inconel (3)	1000	1080	5
Hastelloy (4)	1800	2050	2.5
Alloy Steel	200	200	25
Al-alloy	600	200	24
Lead	150	215	23

Notes:

(1) Information collated from several sources

(2) Quotations fluctuate widely depending on quality specified for material and fabrication, complexity of fabrication, commercial conditions, etc. so data in Table must be treated as only approximate indicators of relative costs

(3) For high Nickel grade of alloy range

(4) For Hastelloy C grades of alloy range

Table 2.4

Properties of Specimen Candidate Filler Materials

A: Low Melting Point Metals and Alloys

Material	Melting Point °C (1)	Density kg/m^3	Remarks (2)
Lead (Pb)	327	11350	Plastic flow behaviour at low stresses. Very good gamma radiation attenuation.
Tin (Sn)	232	7300	About 20 times as expensive per unit volume as Pb.
Zinc (Zn)	419	7130	About twice as expensive per unit volume as Pb.
Aluminium - silicon alloys	about 600	2650	Approximately similar cost per unit volume as Pb

Notes:
(1) Thermal conductivities are high varying from 35 W/m/k for Pb to 120 W/m/k for Zn and more for Al-Si alloys.

(2) Pb and Sn are more cathodic (protected) but Zn and Al-Si more anodic (corroded) than steel in the seawater galvanic series.

Table 2.4 (Cont'd)

B: Miscellaneous Materials

Material	Thermal Conductivity W/m/K	Density kg/m³	Remarks
Cement Grout	1.1	2400	Can be injected at ambient temperatures in fluid state before solidification. Contains Hydrogen and Oxygen in form of water. Includes small voids unless modified by resin additions. Brittle when set.
Two part curing plastics	1.2 to 3.0	900 to 2200	Can be mixed and injected at ambient temperatures in fluid state before solidification. Contains Hydrogen and sometimes Oxygen in molecular structures. Many grades produced with wide range of mechanical properties
Thermoplastics	1.2 to 3.0	900 to 2200	Must be heated (100 to 300°C depending on grade) for injection. Other remarks as for two part curing plastics.
Glass or sand beads	depends on bead shape and size	upto 2000	Includes voids unless matrixed with other filler material
Steel shot	better than glass beads	upto 6500	Includes voids unless matrixed with other filler material
Borosilicate glass	1.1	2500	Must be heated to above 1000°C for injection. Brittle at ambient temperature

Table 2.4

Table 2.5

General Comparison of Containment Options

Containment Designation	Overpack Material (1)	Thickness (mm) (2)	Assembled Wt. (kg) (3)	Relative cost of Overpacking (4)
Corrosion resistant	Titanium alloy	5	2065	100
Corrosion allowance	Steel	75	4000	130
Pressure resistant	Steel	200	9200	280
Radiation Shielding	Steel	300	15400	470

Notes:

(1) Most promising corrosion resistant or corrosion allowance candidate material based on current results of various evaluation programmes. Lead is assumed for void filling.

(2) Estimates of required containment wall thickness for 500 year seabed embedment design life are based on preliminary corrosion behaviour studies so must be considered as conjectural at this stage.

(3) Assembled weight values are for 1.5m internal height, 0.5m internal diameter overpacks and include 500 kg for an HLW canister and 1500 kg for lead void filler.

(4) Relative cost of overpacking indices (corrosion resistant overpack with lead void filler = 100) must be treated as only giving guidance of a most general and approximate nature. It is assumed that titanium alloy overpacks are 17 times as expensive per tonne as steel overpacks to fabricate and lead is 1.5 times as expensive per tonne as steel.

Table 2.7

Galvanic Series of Materials in Seawater

Corrosion resistant construction	GALVANIC SERIES (adapted from references 45 & 46)	Corrosion allowance construction
HGW	Protected (cathodic reaction)	HGW
	Platinum and Gold	
	Graphite	
Overpack	Titanium	
	Hastelloy C	
Canister	Stainless Steel (passive)	Canister
	Inconel, Monel and Copper	
	Tin	
Filler	Lead	Filler
	Stainless Steel (active)	
Penetrator Body	Cast Iron and Steels	Penetrator Body
	Aluminium	
Galvanising	Zinc	Galvanising
SEABED	Corroded (anodic reaction)	SEABED

Table 2.6

General Comparison of Penetrator Construction Options

Containment Designation	Overpack Details	Penetrator Tube Details	Assembled Wt. per canister (kg) (3)	Relative Cost of Penetrator Construction (4)
Corrosion resistant	5mm wall Ti-alloy (1)	25mm wall steel	3600	100
Corrosion allowance	none (2)	75mm wall steel	4600	100
Pressure resistant	none (2)	200mm wall steel	8200	175
Radiation shielding	none (2)	300mm wall steel	11900	250

Notes:

1) Assumes HLW canister contained by 5mm wall thickness titanium alloy overpack (engineered barrier) inside 25mm wall thickness penetrator tube (to resist handling and seabed impact loads, etc.)

(2) Assumes penetrator tube also acts as engineered barrier.

(3) Assembled weight values are approximate. In addition to the overpack and 1.5m long 0.5m internal diameter penetrator tube weight, they include 500 kg for an HLW canister, 1500 kg for lead void filler and 1000 kg allowance for penetrator nose and tail pieces.

(4) Relative cost of penetrator construction indices (corrosion resistant option = 100) must be treated as only giving guidance of a most general and approximate nature. They include the Ti-alloy overpack, steel penetrator tube, nose and tail and lead void filler assuming Ti is 17 times as expensive as steel and lead is 1.5 times as expensive per tonne as steel.

Table 6.1

Penetrator Tube Manufacturing Length Limits

Process	Min. Wall Thickness	Max. Manufacturing lengths (1) (2)		
		25mm wall	75mm wall	200mm wall
Forging or Solid casting + machining	depends on length	8m	more than 20m	more than 20m
Hollow casting	depends on length	wall too thin	7.5m	7.5m
Centrifugal casting (3)	10mm	9m	9m	3.5m
Extrusion (4)	20mm	12m	7m	2.5m
Plate bending & seam welding	no min.	12m (5)	about 40mm wall thickness limit	
Pipe milling	6mm	more than 20m (6)	more than 20m (6)	about 100mm wall thickness limit

Notes:

(1) Applicable to 500 mm internal diameter tubes

(2) Based on capacity of current industrial equipment and methods

(3) Japanese companies reported to have equipment for up to 50t tubes but centrifugal casting length limits given for current European capacity for up to 12 t pipes not exceeding 9m length

(4) Extrusion length limits given for current European facilities but equipment for extruding 25% longer pipes installed in America

(5) Typical length of large industrial "U" - "O" presses

(6) Assuming special arrangement for supply of non standard long pipe lengths

Table 6.2

Evaluation of Penetrator Tube Manufacturing Options

Process	Tube length (mm)	1500			6500			15600		
	Tube wall th. (mm)	25	75	200	25	75	200	25	75	200
Solid forging + machining		3=	3=	3=	3=	3=	3=	0	3=	3=
Hollow forging + machining		2	2	2	2	2	2	0	0	0
Solid casting + machining		3=	3	3=	3=	3=	3=	0	3=	3=
Hollow casting		2	2	2	2	2	2	0	0	0
Centrifugal casting		3=	3=	3=	3=	3=	1	1	1	1
Pipe extrusion		3=	3=	3=	3=	3=	1	1	1	1
"U"-"O" pressing and Seam welding		3=	0	0	3=	0	0	1	0	0
Plate rolling and Seam welding		3=	0	0	0	0	0	0	0	0
Pipe milling		2	2	1	2	2	1	2	1	1

Key:

0 - not feasible
1 - feasible but outside current manufacturing capacity
2 - within current manufacturing capacity
3 - most appropriate process (if any)

Table 6.3

Evaluation of Circumferential Welding Options

Process	Steel Wall Thickness (mm)					
	25		75		200	
	Fab	Ass	Fab	Ass	Fab	Ass
Arc welding	4=	1	2	1	2	1
Narrow gap arc welding	2	1	4=	3=	4=	3=
Electron beam welding (EBW)	4=	4=	4=	4	4=	4
Laser welding (LW)	1	3=	0	0	0	0
Plasma welding (PW)	0	0	0	0	0	0
Friction welding (FW)	4=	4=	1	3=	0	0
Radial friction welding (RW)	1	3=	0	0	0	0
Diffusion bonding (DB)	3	3=	3	3=	3	3=

Fab: Fabrication welding in conventional steelworks
Ass: Assembly welding by automated processes
Key: 0 - not feasible or not envisaged in near future
1 - under development for industrial application in near future
2 - within current industrial welding capacity
3 - most appropriate process under development (if any)
4 - most appropriate current process (if any)

Table 6.4

Quality Assurance and Quality Control (QAQC) During Fabrication

Fabrication Stage (2)	Check Points (3)
Examination of plans and calculations	6
Receipt and control of parent materials	20
Receipt and control of welding consumables	20
Qualification of welding procedures	30
Qualification of welders	23
Control of work preparation before welding	4
Control of work preparation during welding	15
Control of work preparation after welding	20
Control of heat treatment	20
Performance of final tests including NDT	6

Notes:

(1) International Institute of Welding Commission XI Working Group on Quality Assurance recommendations (65).

(2) Pressure vessel construction example.

(3) Total of 164 check points allocated over fabrication sequence.

Table 7.1

Comparison of Nominal 7,500t Displacement Ship Design Options

Ship Option	Length[1] (m)	Breadth (m)	Approx. Reference Design Penetrator Capacity[2]
A	100	18	75
B	130[3]	16.5	60
C	100	20	72
D	105	16.5	46
E	100	20	18

Note (1) All ship details are approximate and necessarily conjectural at this early development stage. The ship lengths given are, in particular, only indicative. Fuller consideration of ship draught constraints and sea handling characteristics would be necessary to identify the most appropriate proportions. Typical draughts would be expected to lie between 5 and 6m.

(2) Ship designs B, C and D as shown in Figures 39 to 41 are arranged for penetrators other than the reference design. Equivalent numbers of reference penetrators are given in the Table in these cases.

(3) Ship option B would have a nominal displacement of 10,000 t. See Section 7.6.3.

Table 7.2

Disposal Voyage Cycles (1)

DISPOSAL SITE (See Figure 1)	GME	CV2	SNAP
Sailing Distance (km)	3000	4500	6200
IN PORT:- (days) Provisioning and Penetrator Loading	12	12	12
AT SEA:- (days) Outward Voyage (1)	7	10	14
Penetrator Release (2)	6	4	6
Return Voyage (1)	7	10	14
TOTAL (days)	32	36	46
Maximum Number of Voyages per Year per Vessel (3)	10	9	7

Note (1): Operational details such as sailing speeds and times for penetrator loading and release are given for illustration only and are necessarily conjectural at this early development stage.

(2): Assuming 6 m/s (12 knots) ship speed.

(3): Assuming 2 day weather windows for penetrator release operations occur on average one in any 6 days at GME and SNAP and once in any 4 days at CV2.

(4): 2 months (15%) of year allowance included for ship annual maintenance and survey and contingencies.

Table 9.1

Disposal Option Cost Summary

	Reference Option	Conservative Option
Initial capital cost	£129m	£135m
Annual operating cost	£10.7m	£22.1
Combined cost on an annual basis	£26.9m	£39.1m
Cost per HLW canister	£47,100	£68,600
Cost per MW(e)yr	£600	£880
Cost per kW(e)hr	0.007p	0.010p

Note: (1): See text for description of operations and assumptions made for cost assessments.

(2): Costs estimated at March 1984 price rates.

Table 9.2

Cost Breakdown

	Reference Option			Conservative Option		
	No.	Unit Cost	Total Cost	No.	Unit Cost	Total Cost
Capital Items		£ x 10^3	£ x 10^6		£ x 10^3	£ x 10^6
(a) 12½ year amortization:-						
Canister Flasks	6	750	4.5	6	750	4.5
Penetrator Flasks	2	1000	2.0	5	600	3.0
Railway Wagons	6	50	0.3	6	50	0.3
Assembly Equipment	-	-	32.0	-	-	32.0
Dock Equipment	-	-	5.0	-	-	3.0
Disposal Ships	1	-	15.0	1	-	22.0
Total Cost			58.8			64.8
Annual Cost			8.4			9.3
(b) 25 year amortization:-						
Assembly Buildings	1	-	60.0	1	-	60.0
Dock Civil Works	1	-	10.0	1	-	10.0
Total Cost			70.0			70.0
Annual Cost			7.7			7.7
Operating Items						
Canister Train Journeys	12	8	0.1	12	8	0.1
Penetrator Assembly	115	25	2.9	285	45	12.8
Penetrator Assembly;						
material	115	5	0.6	285	2	0.6
operation	-	-	2.0	-	-	2.0
staff	100	35	3.5	130	35	4.6
Dock Facilities;						
operation	-	-	0.3	-	-	0.3
staff	20	35	0.7	20	35	0.7
Disposal Ships;						
operation	2	100	0.2	3	200	0.6
salaries	10	40	0.4	10	40	0.4
Annual Cost			10.7			22.1

Note Total costs for each option are summarised on Table 9.1

Table 9.2

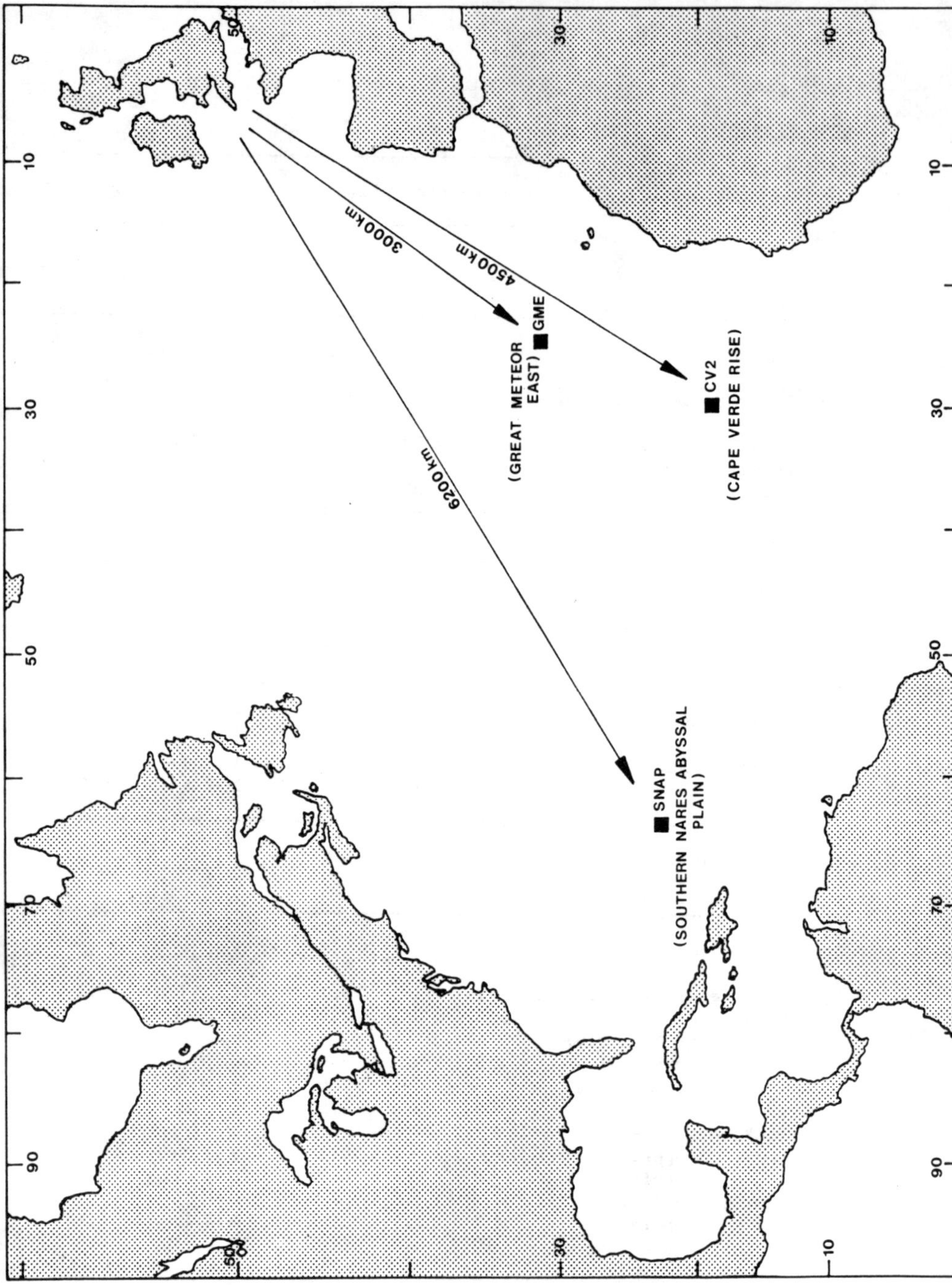

SEABED DISPOSAL STUDY AREAS **figure 1**

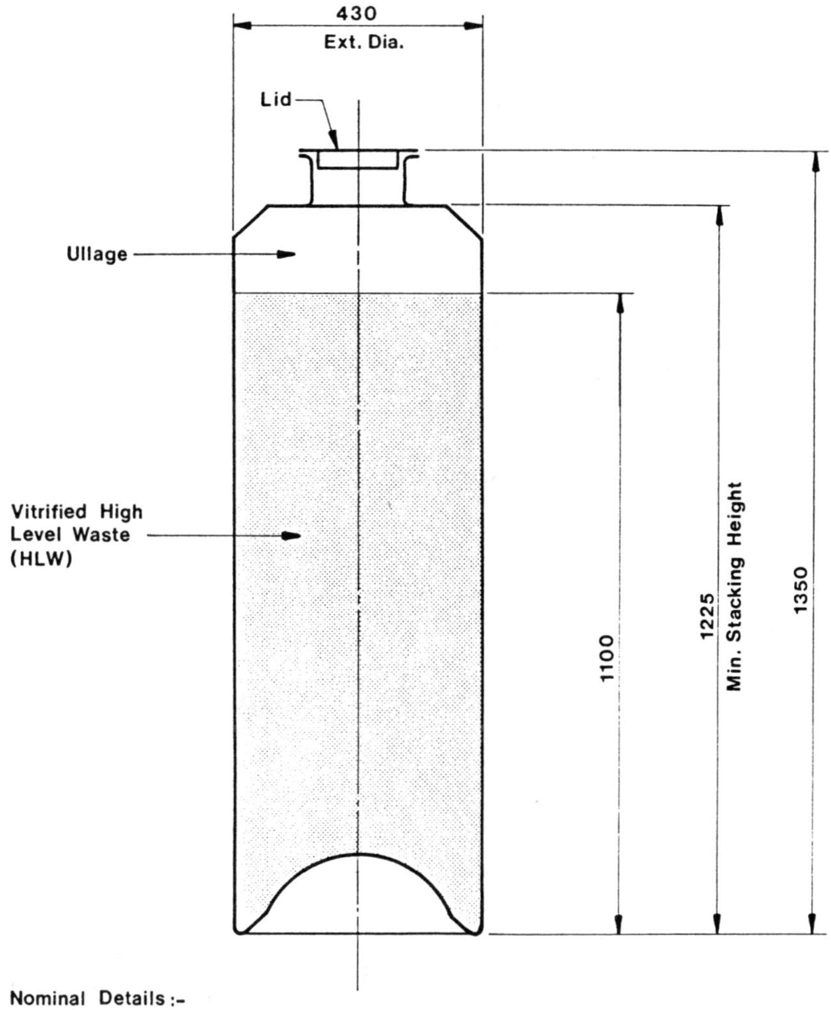

Nominal Details :-

Canister Wall Thickness	5 mm
HLW Volume	0·150 m^3
Ullage Volume	0·030 m^3
Weight when empty	75 kg
Weight including HLW	480 kg
Weight if Ullage filled with Lead	820 kg

HLW CANISTER

figure 2

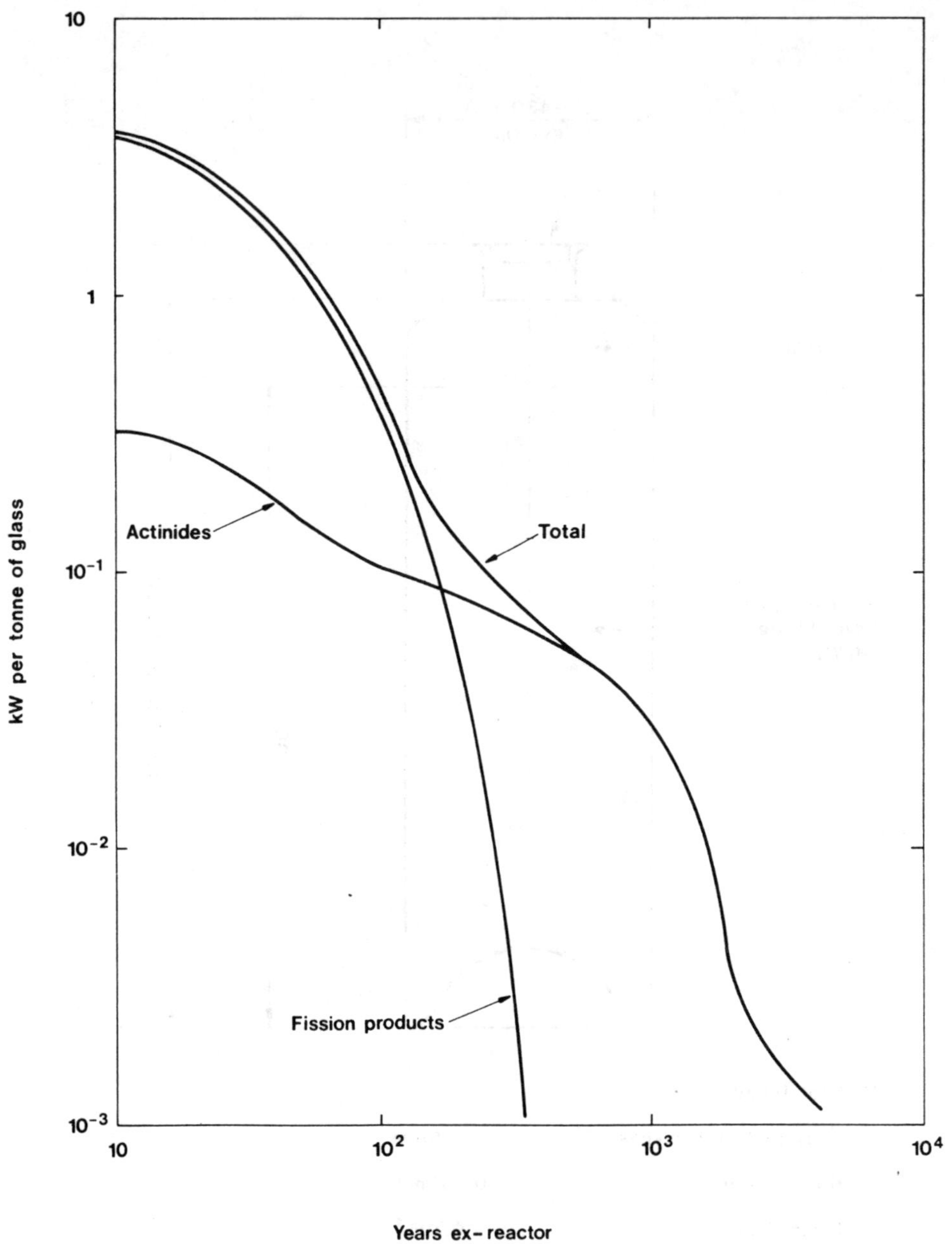

Based on Ref. EUR 8179 for LWR Fuel

HLW HEAT GENERATION DECAY

figure 3

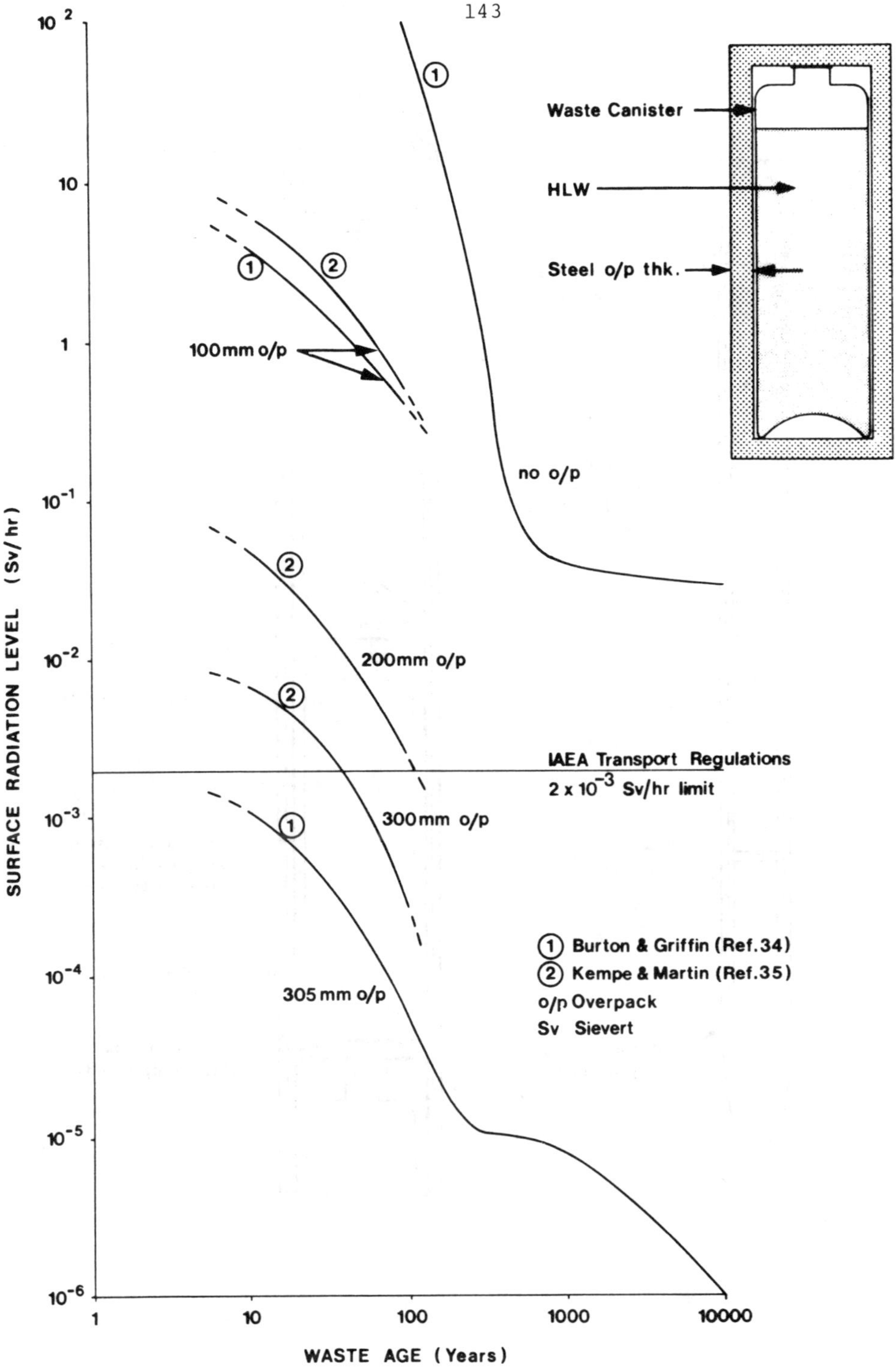

OVERPACK SURFACE RADIATION LEVELS **figure 4**

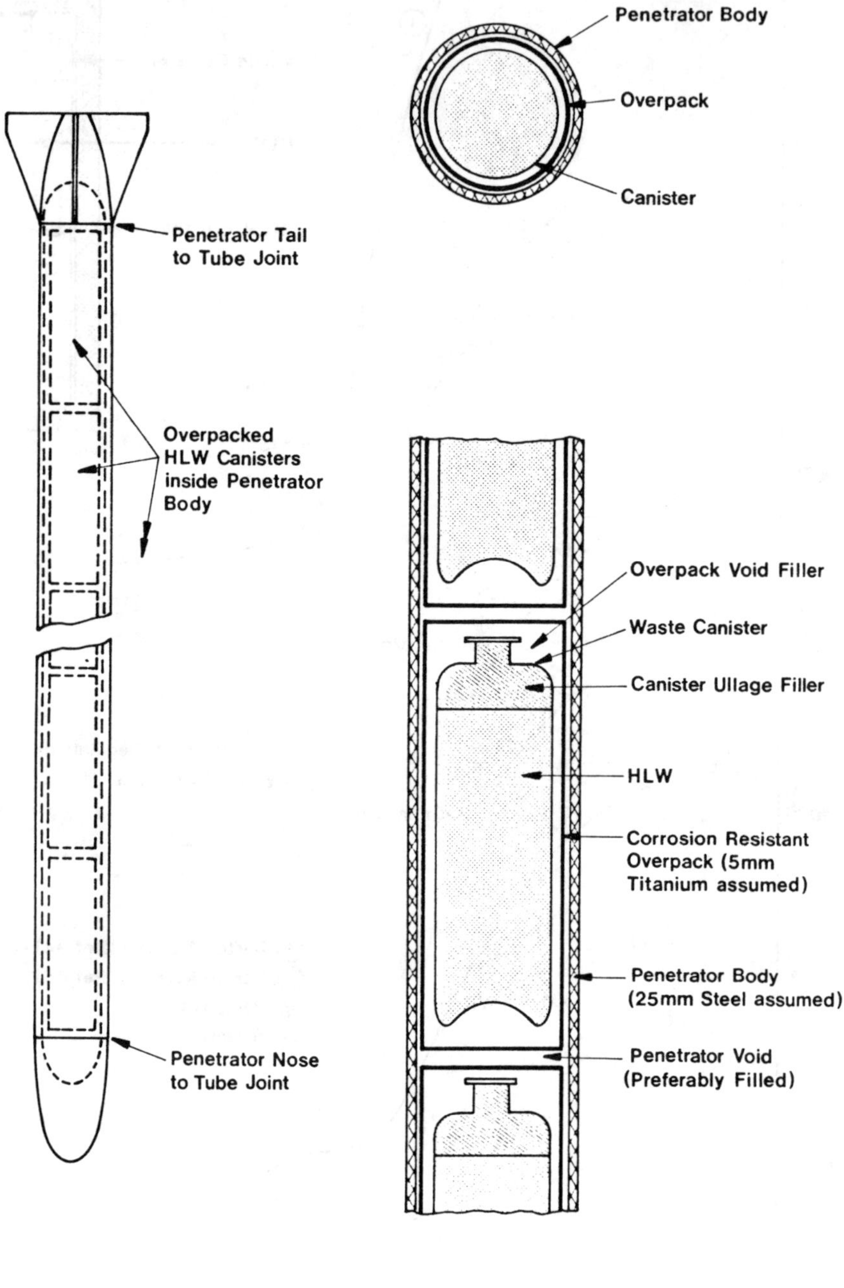

CORROSION RESISTANT CONTAINMENT figure 5

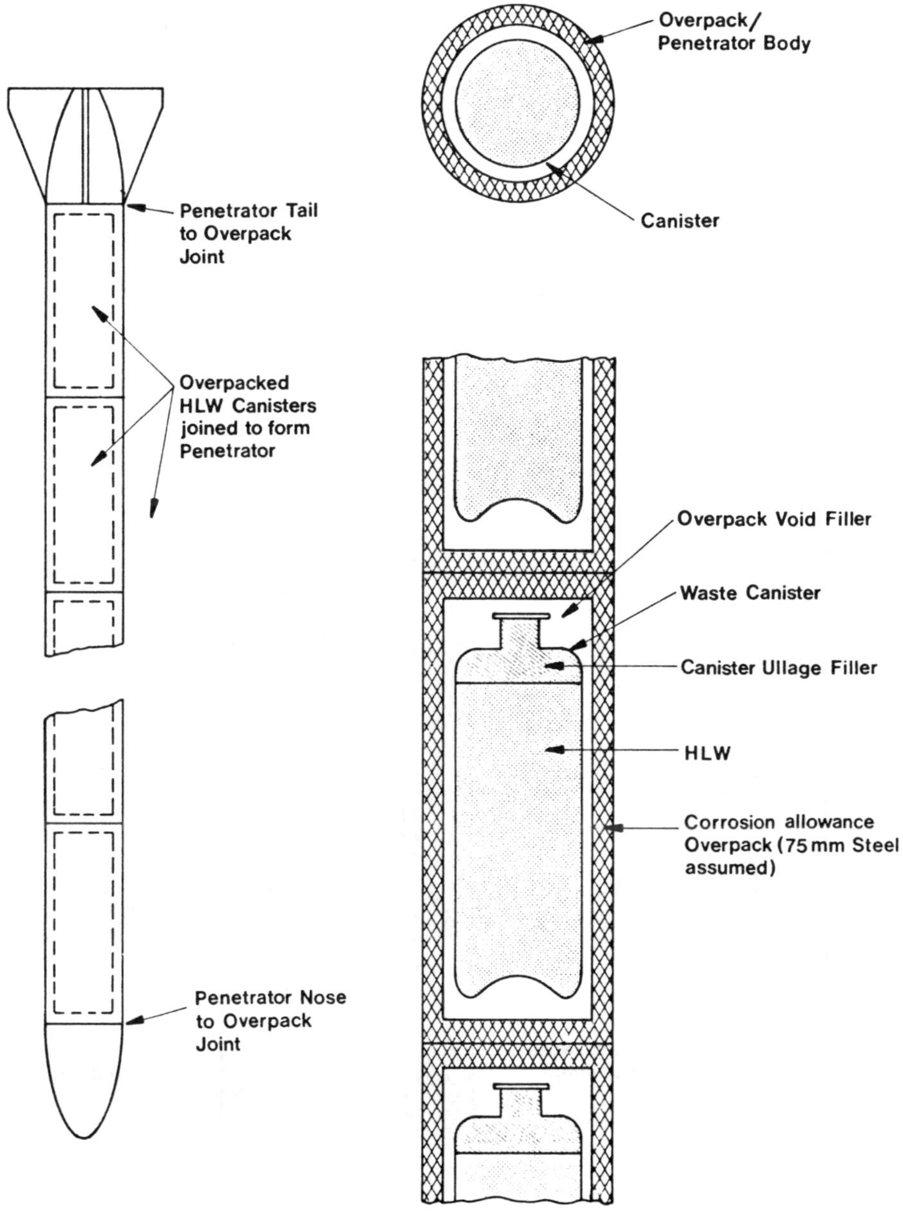

CORROSION ALLOWANCE CONTAINMENT figure 6

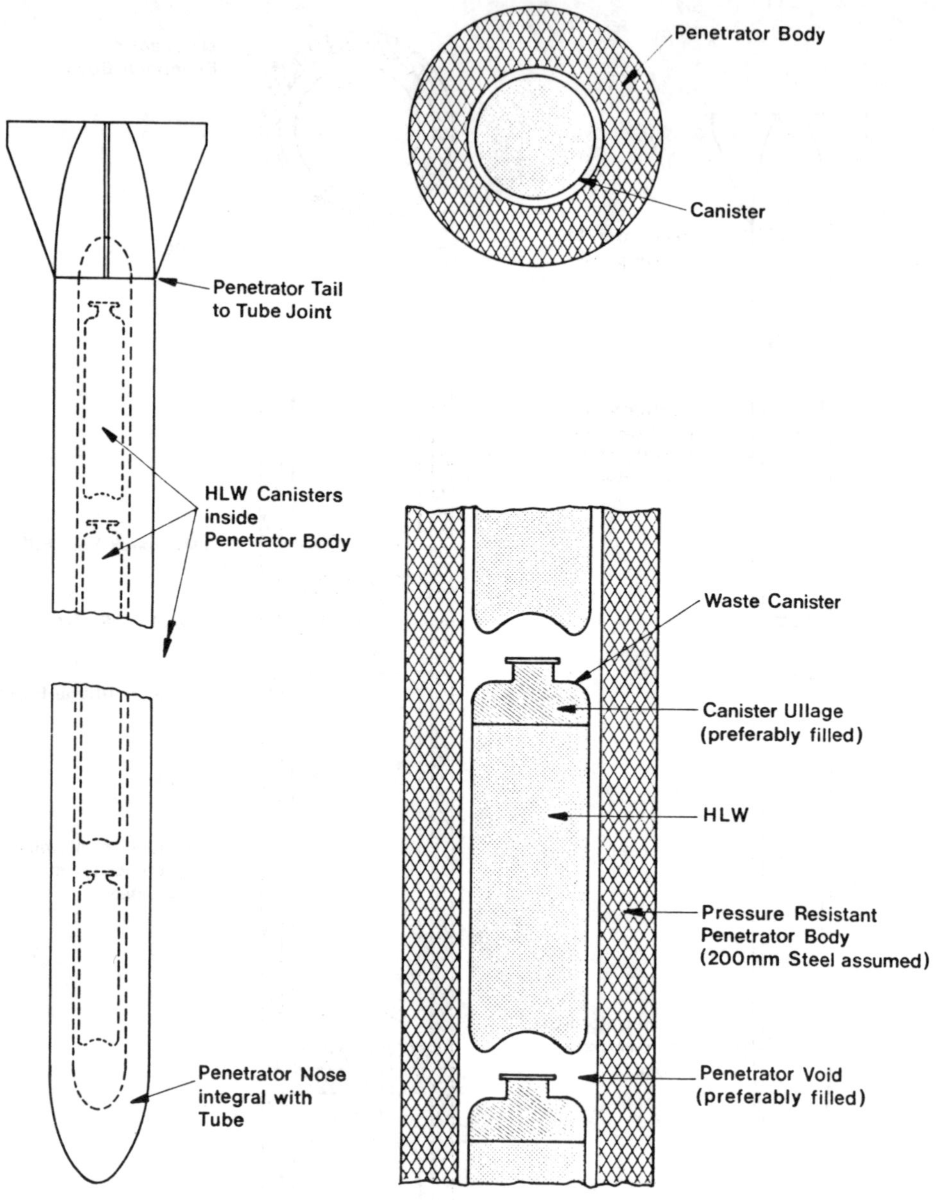

PRESSURE RESISTANT CONTAINMENT figure 7

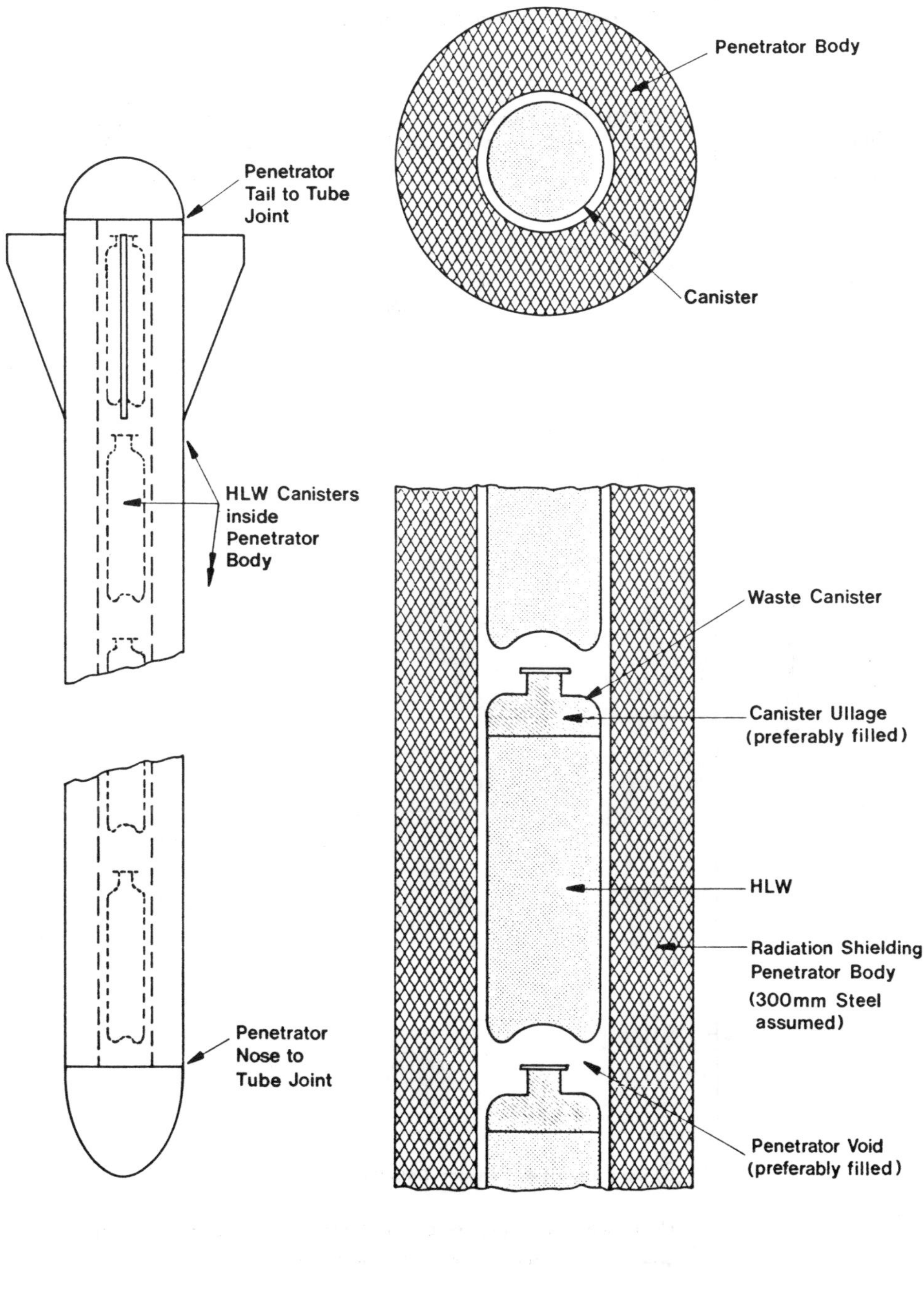

RADIATION SHIELDING CONTAINMENT

figure 8

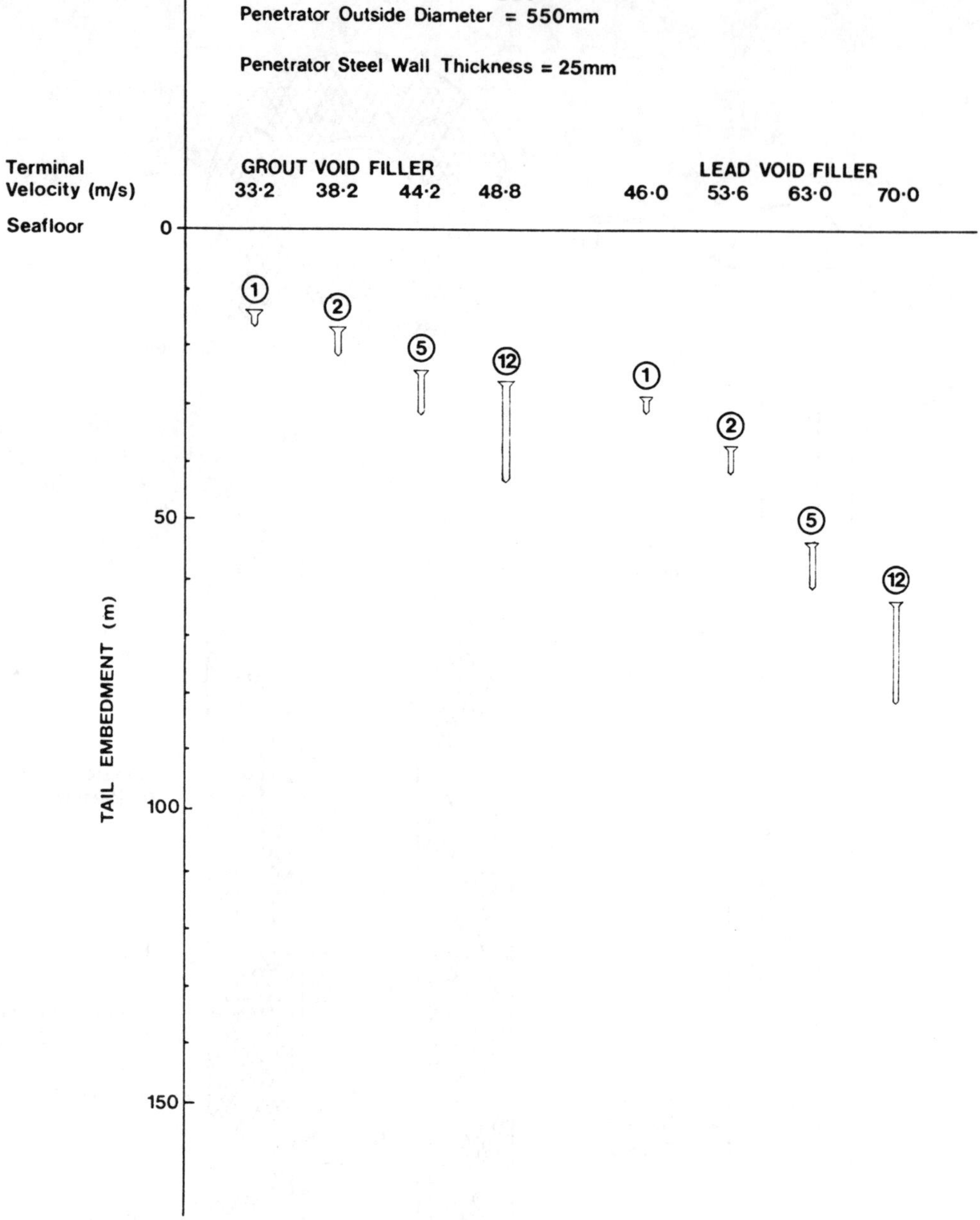

(5) indicates HLW Canister Capacity

Undrained Shear Strength, C_u, of Sediment taken as $5 + 1{\cdot}5z$ kN/m^2
where z is depth in metres below Seafloor

SEABED EMBEDMENT:
CORROSION RESISTANT OPTION

figure 9

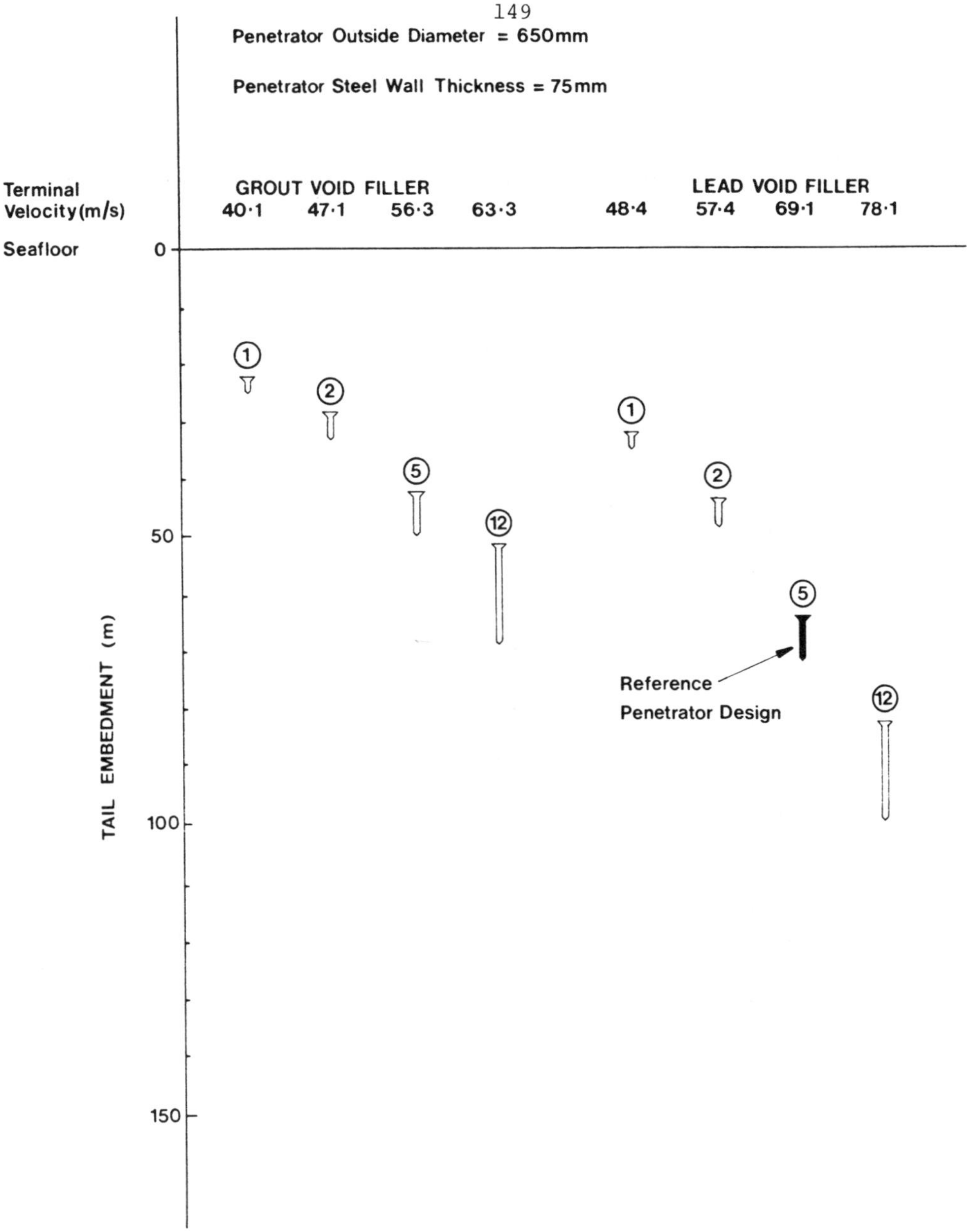

(5) indicates HLW Canister Capacity

Undrained Shear Strength, C_u, of Sediment taken as $5 + 1{\cdot}5z$ kN/m^2 where z is depth in metres below Seafloor

SEABED EMBEDMENT :
CORROSION ALLOWANCE OPTION

figure 10

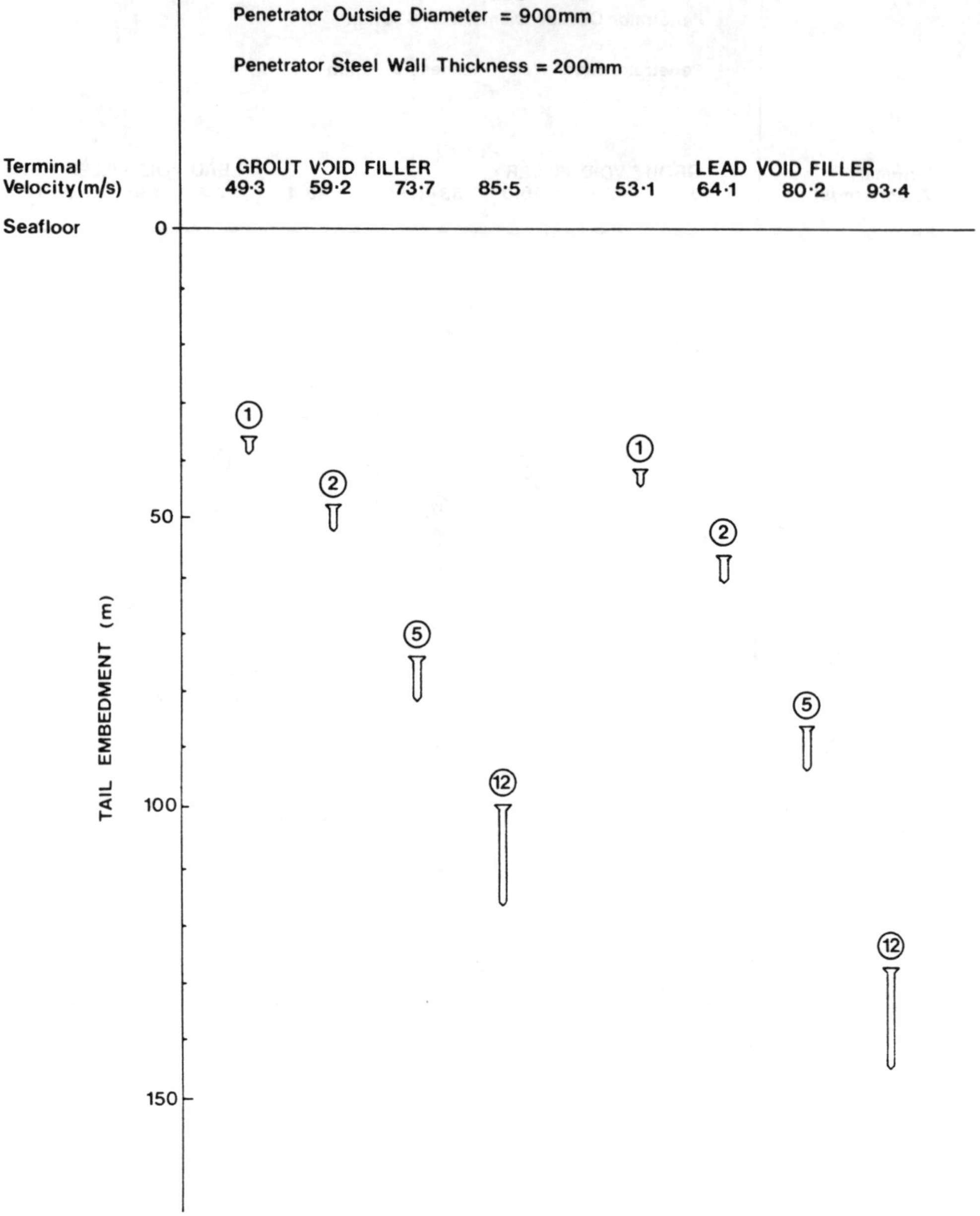

(5) indicates HLW Canister Capacity

Undrained Shear Strength, C_u, of Sediment taken as $5 + 1{\cdot}5z$ kN/m^2 where z is depth in metres below Seafloor

SEABED EMBEDMENT : PRESSURE RESISTANT OPTION figure **11**

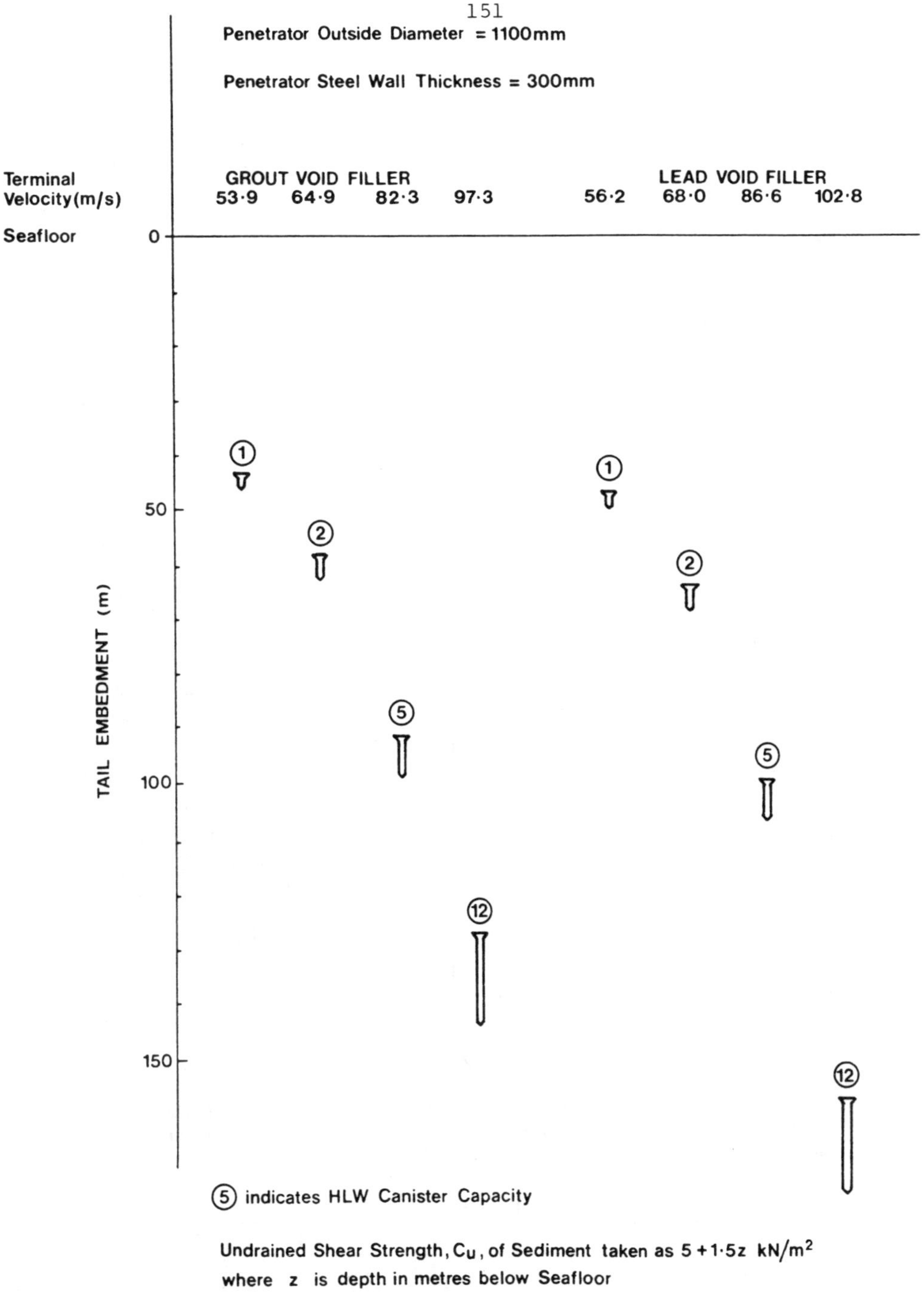

SEABED EMBEDMENT :
RADIATION SHIELDING OPTION

figure 12

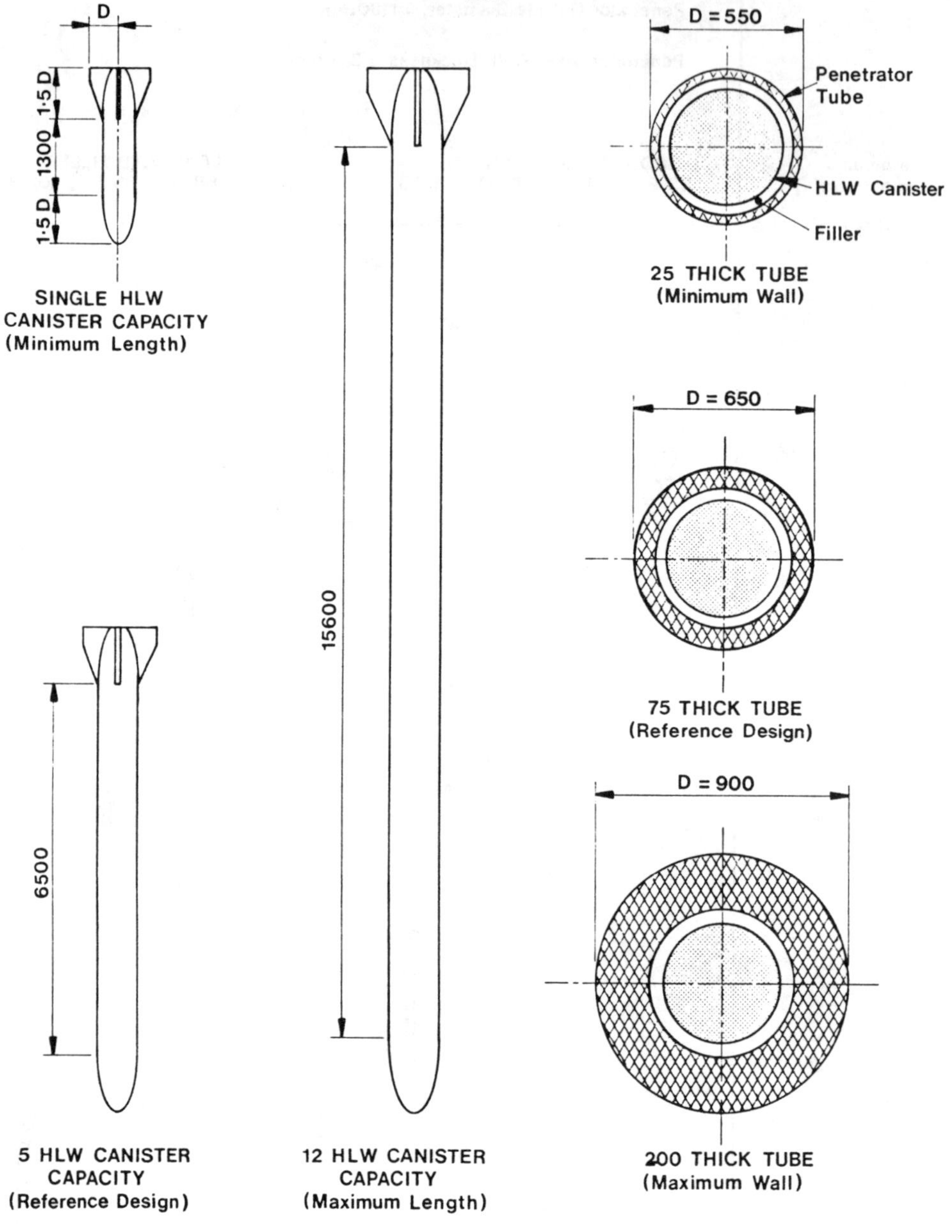

CANISTER CAPACITY		1	2	3	4	5	6	9	12	No.
Approx. Overall Length		3·5	4·5	6·0	7·0	8·5	10·0	13·5	17·5	m
Approx. Total Weight:	D = 550	4	6	8	10	12	14	20	26	
(for steel tube and	D = 650	7	9	12	15	18	21	30	39	t
lead filler)	D = 900	15	21	27	33	39	45	63	81	

PENETRATOR DESIGN OPTIONS figure 13

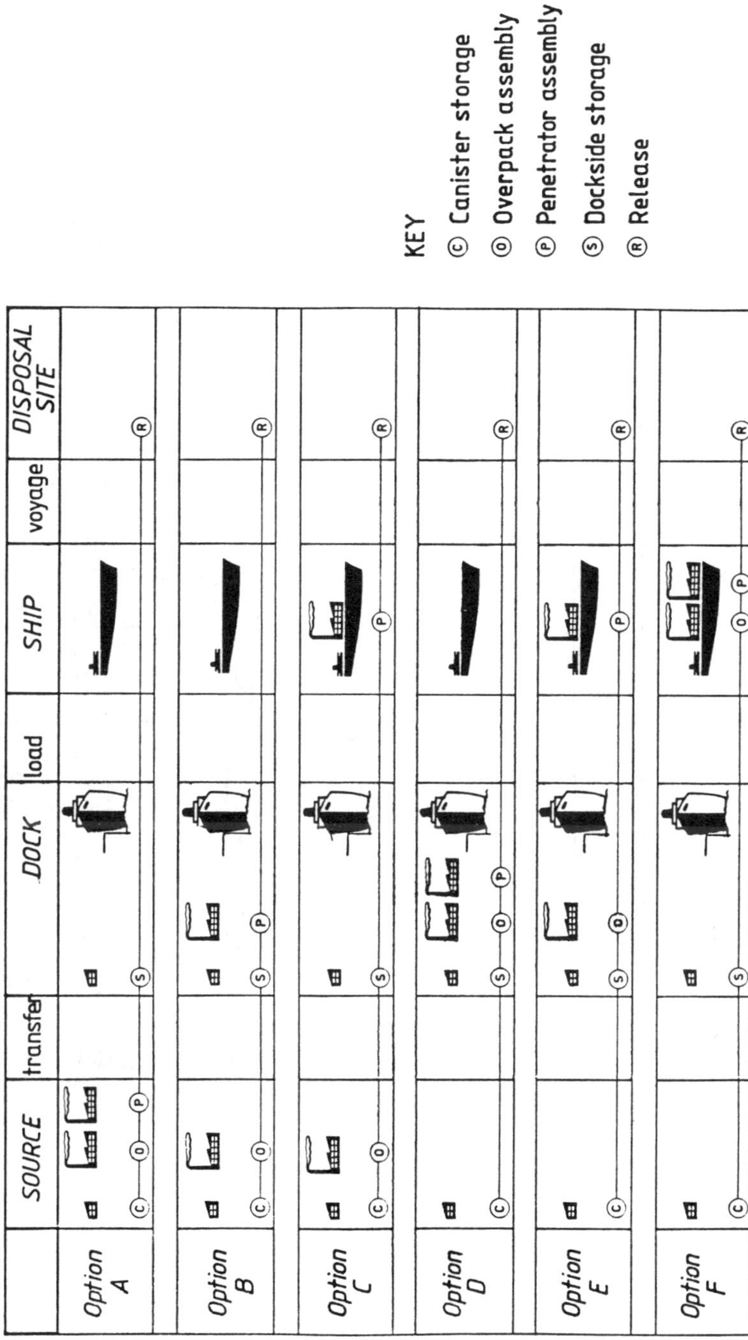

DISPOSAL OPERATION OPTIONS

figure **14**

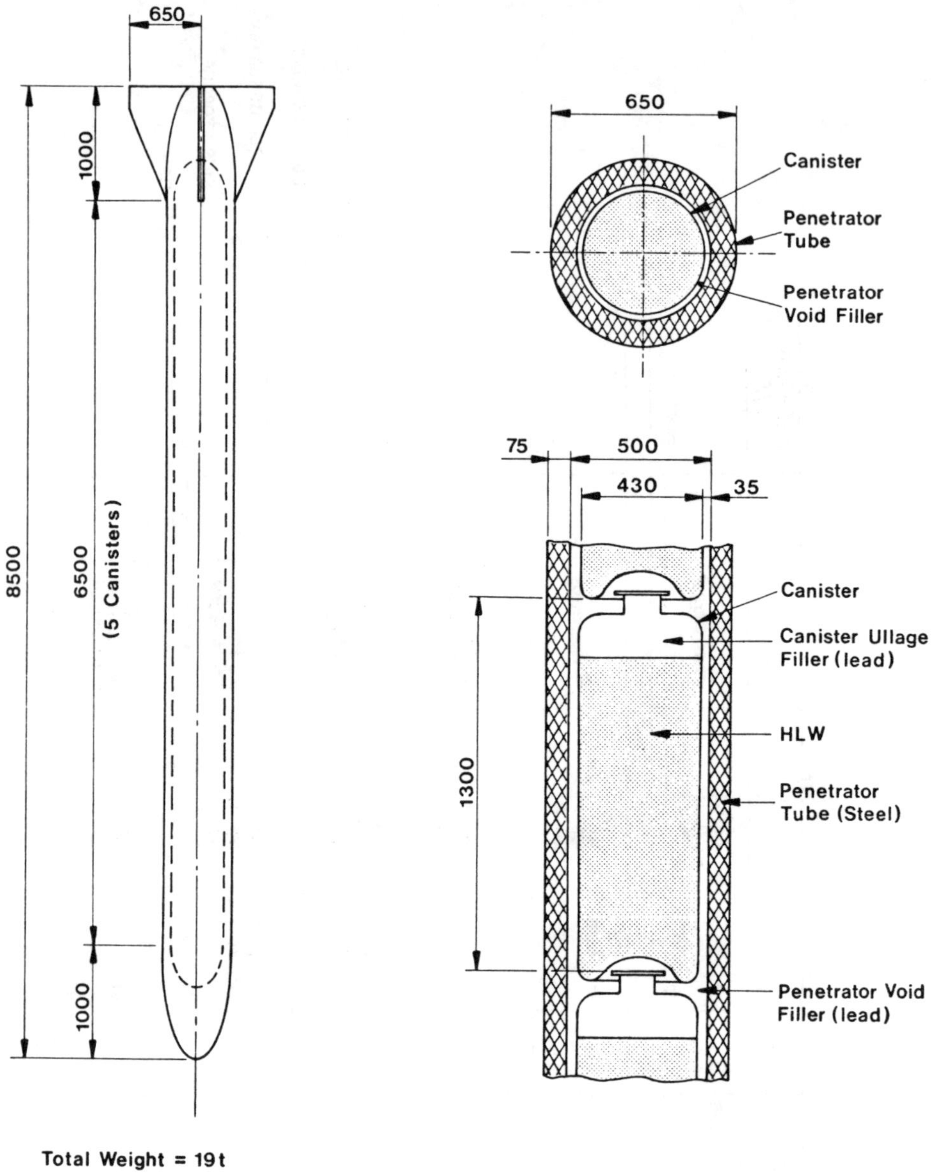

PENETRATOR REFERENCE DESIGN

figure 15

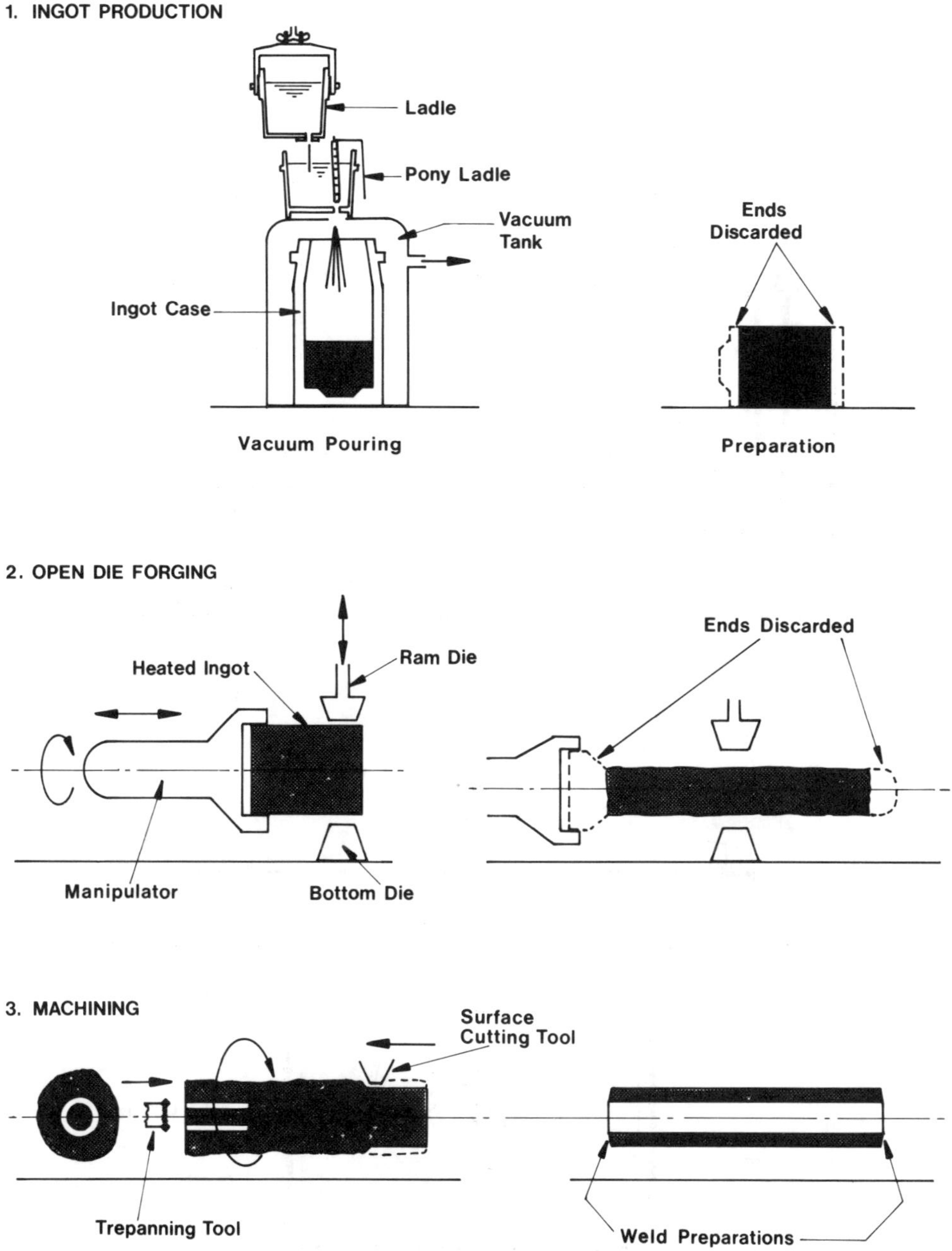

TUBE MANUFACTURE: FORGING figure 16

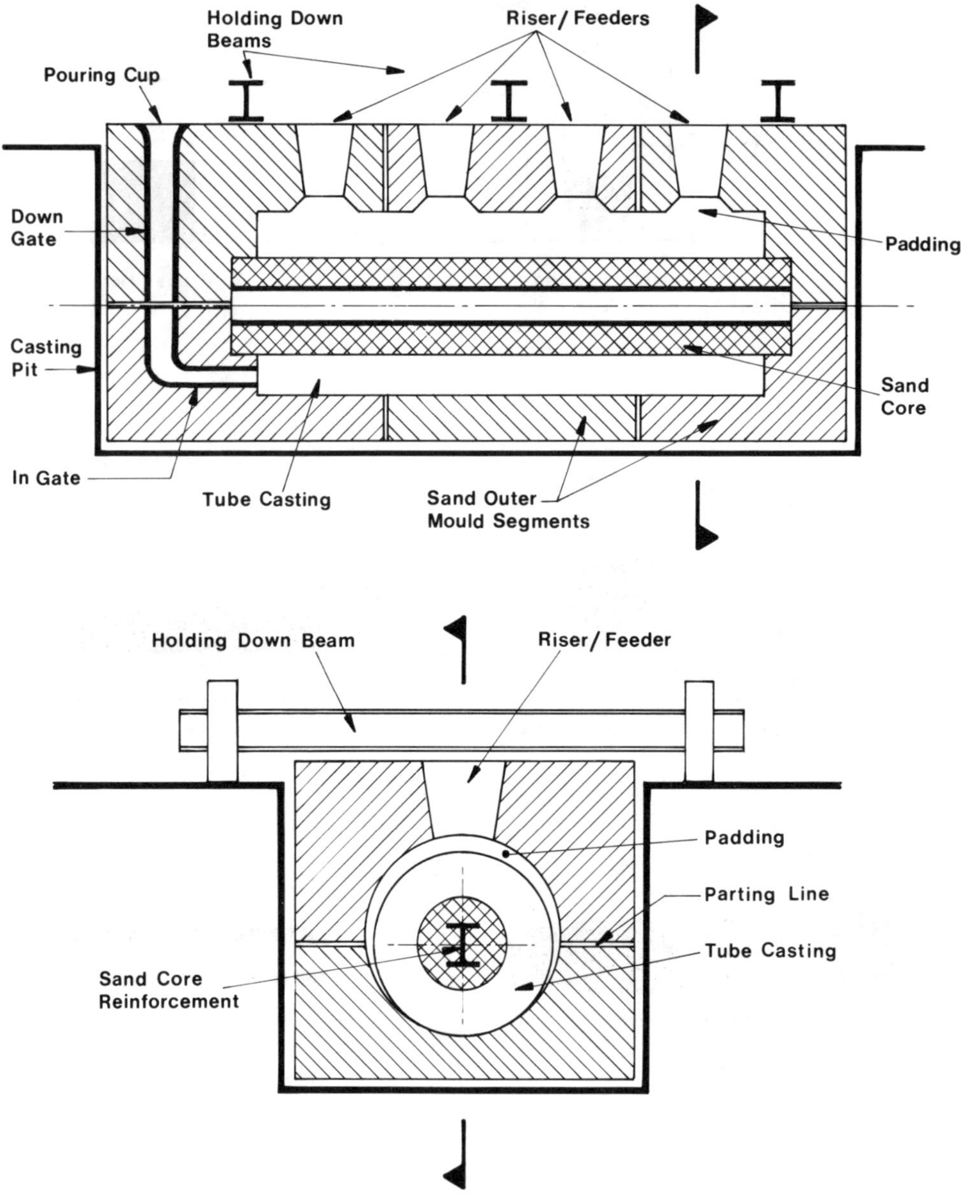

TUBE MANUFACTURE: SAND CASTING

figure 17

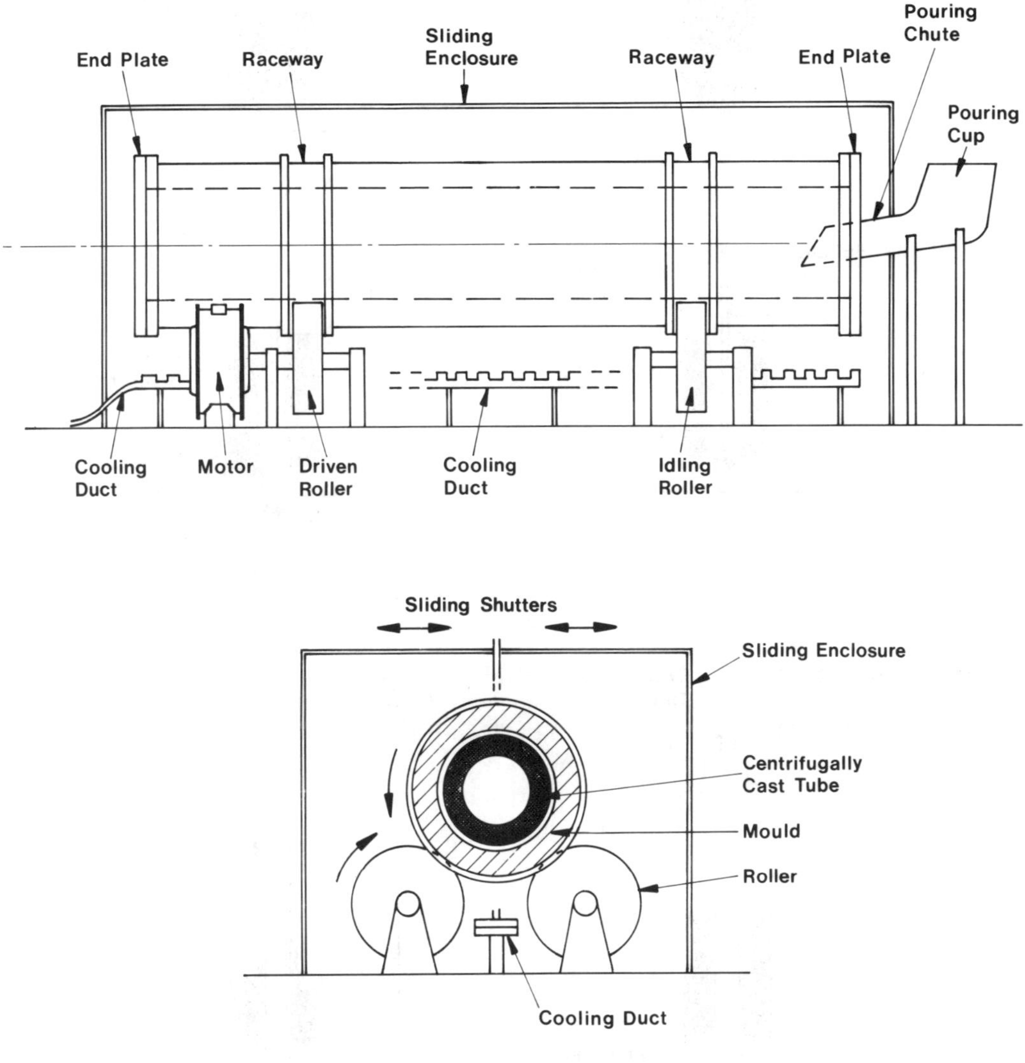

TUBE MANUFACTURE: CENTRIFUGAL CASTING

figure 18

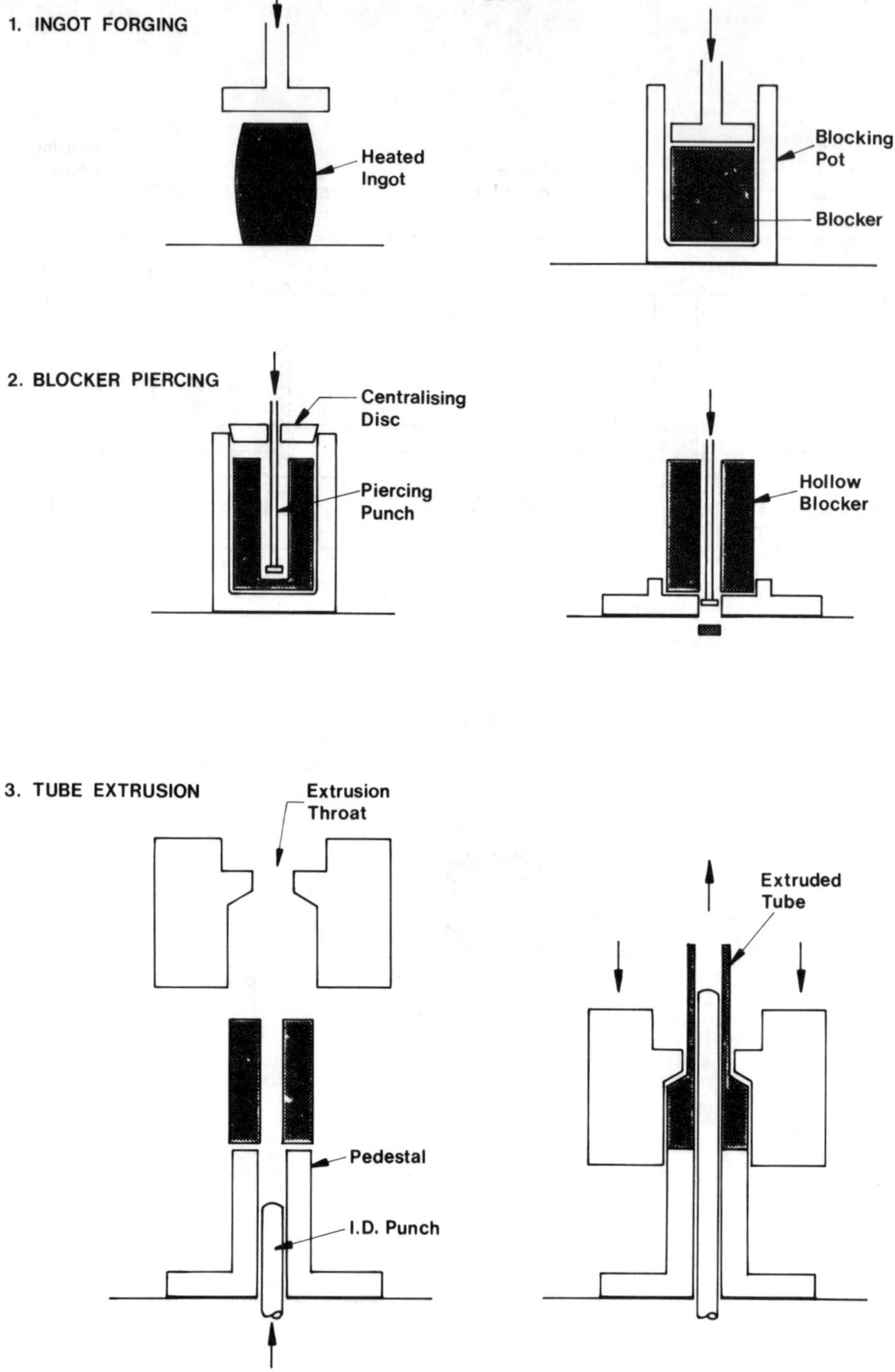

TUBE MANUFACTURE: EXTRUSION **figure 19**

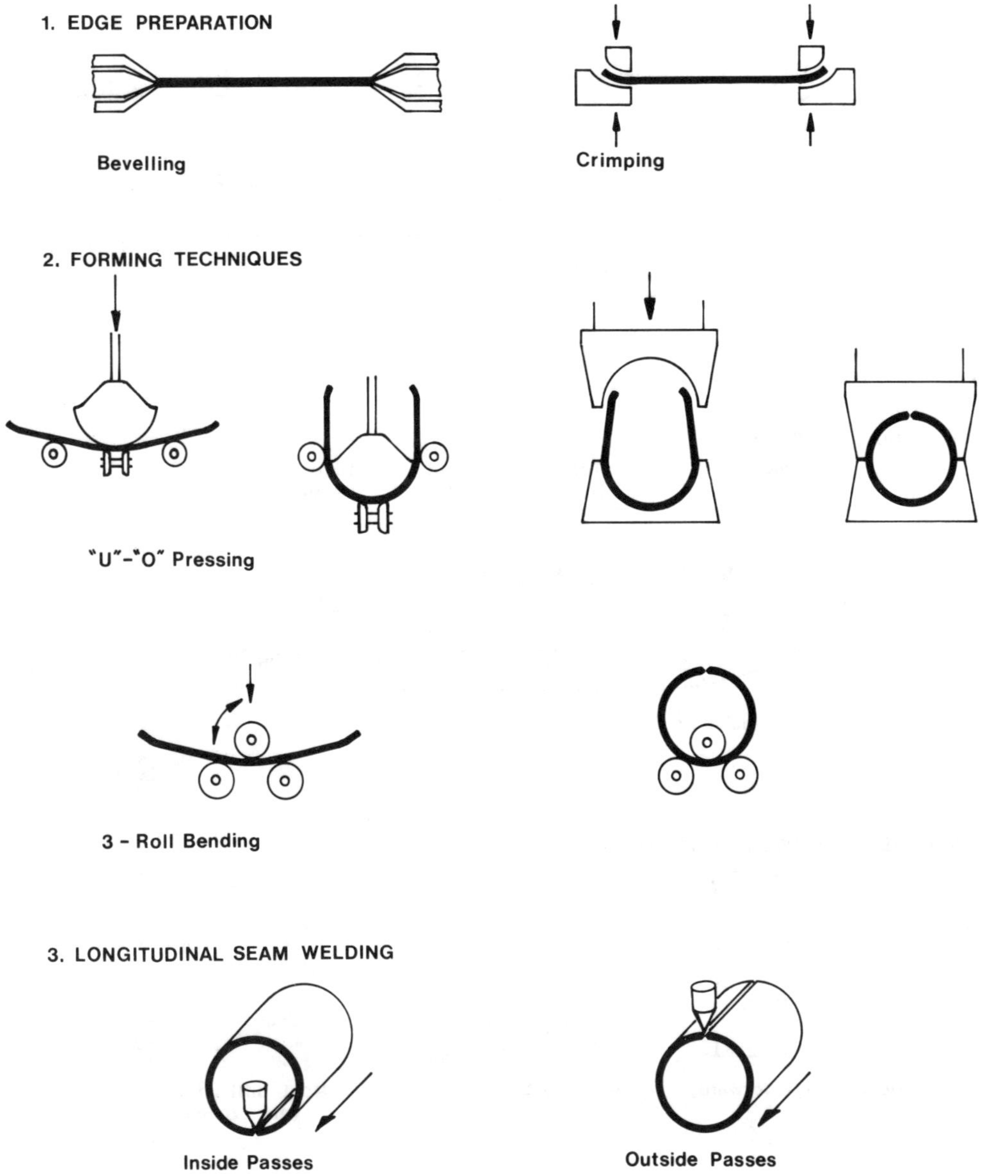

TUBE MANUFACTURE : PLATE FORMING figure 20

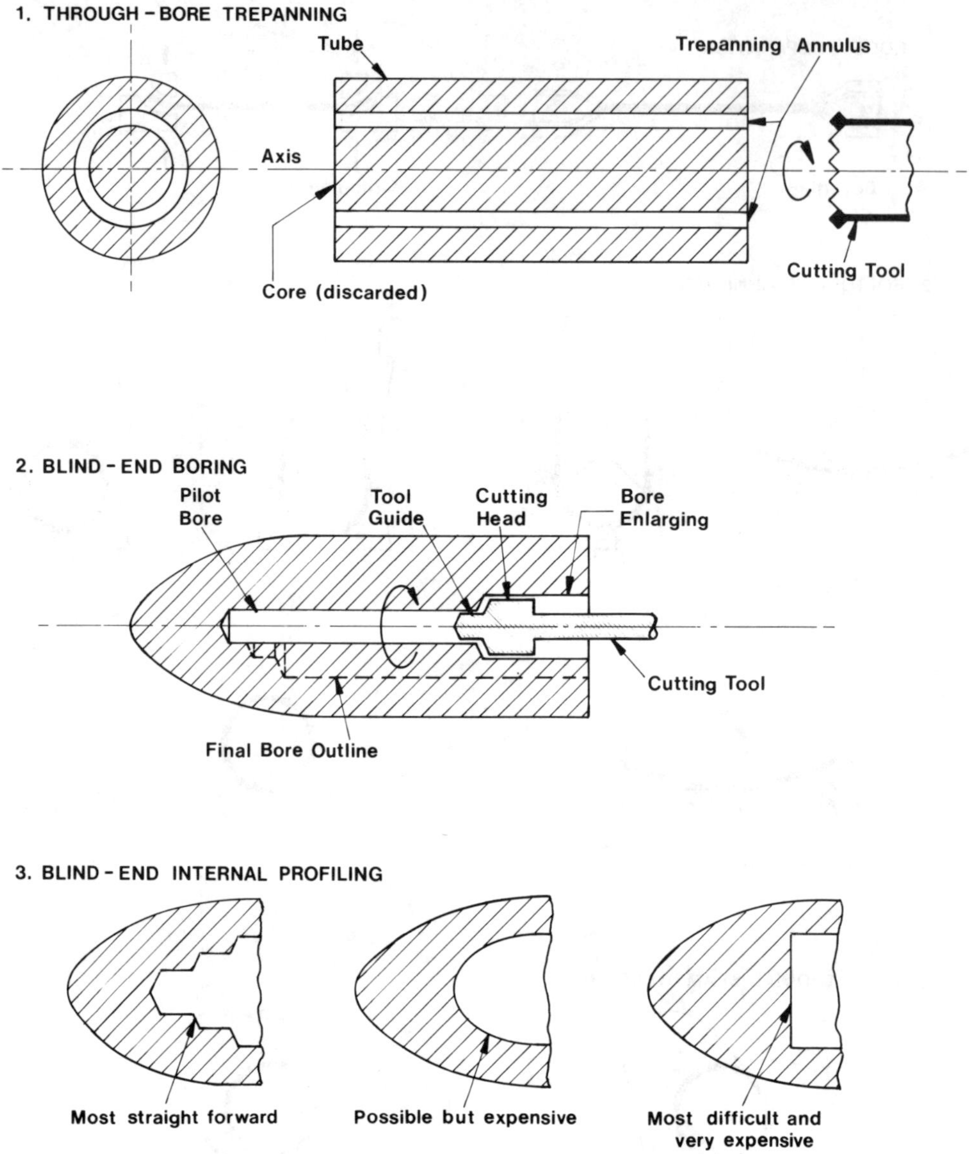

TUBE MANUFACTURE: MACHINING **figure 21**

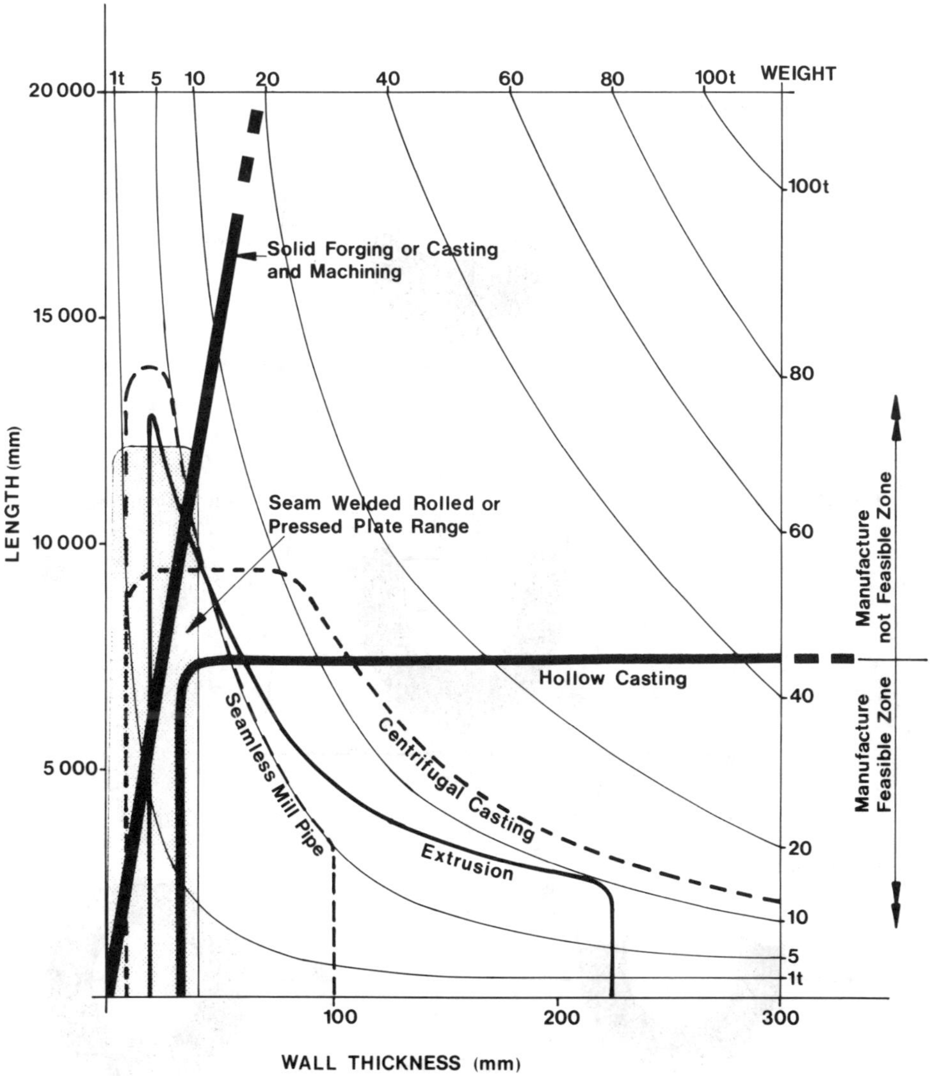

N.B. Ranges apply to 500 mm Internal Diameter Tubes

TUBE MANUFACTURE : TYPICAL SIZE RANGES figure 22

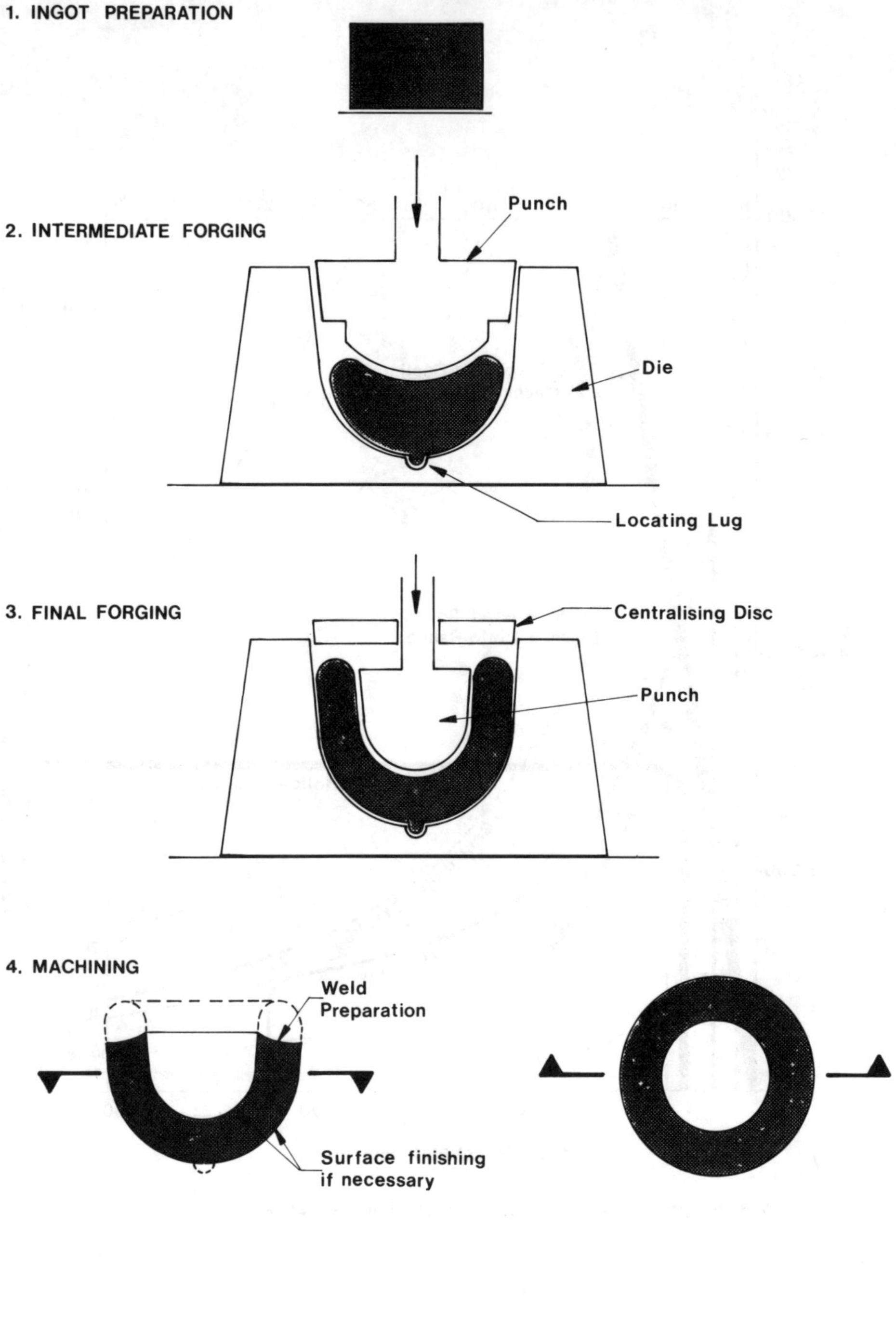

NOSE MANUFACTURE : FORGING

figure 23

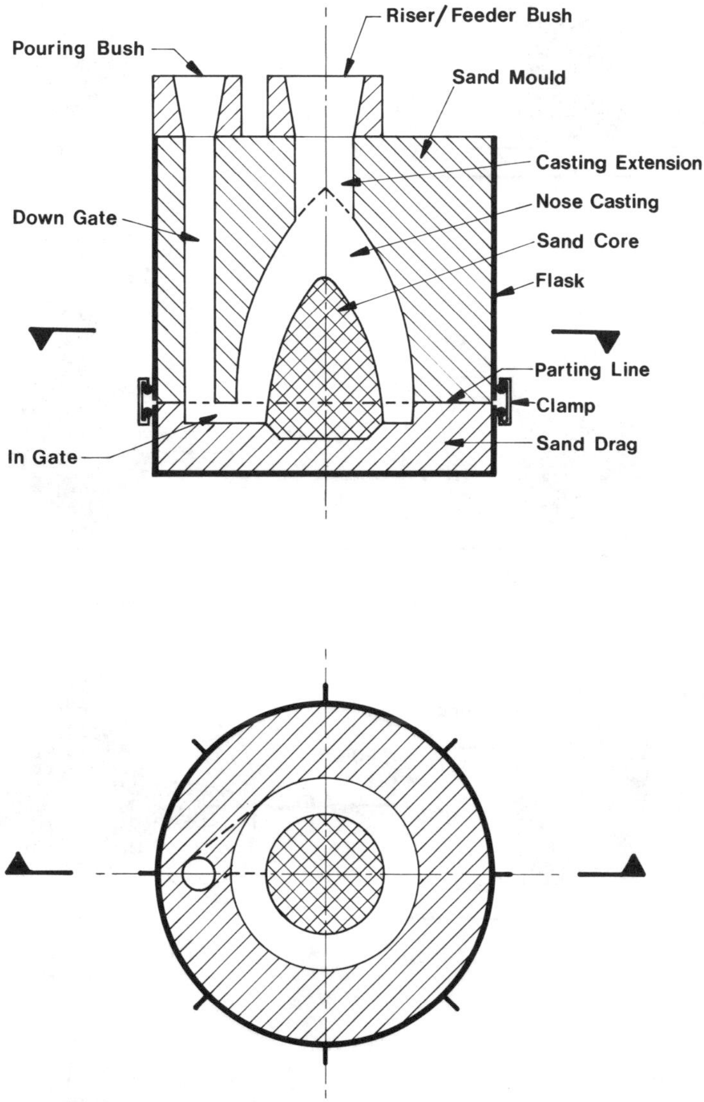

NOSE MANUFACTURE: SAND CASTING figure 24

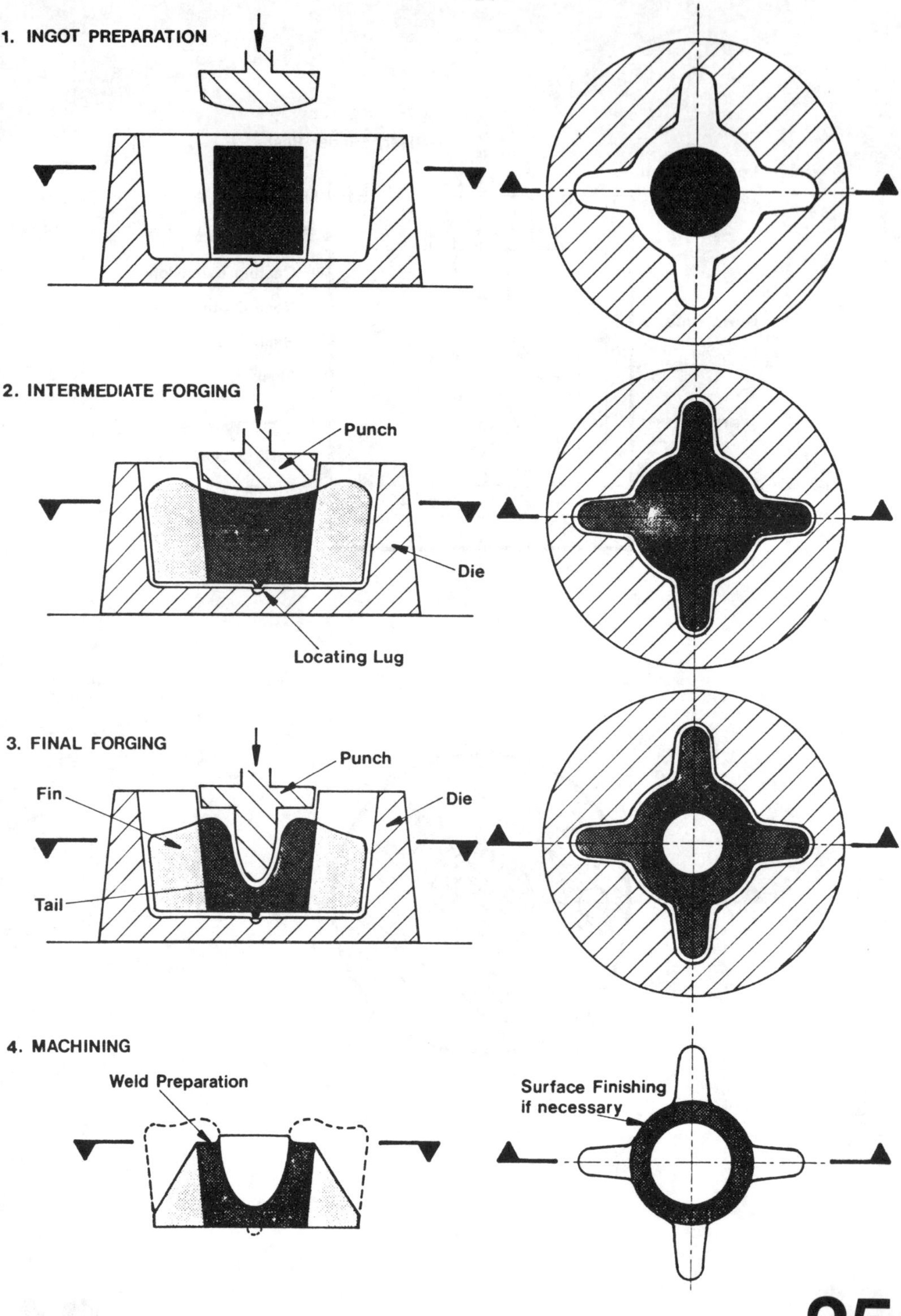

TAIL MANUFACTURE: FORGING

figure 25

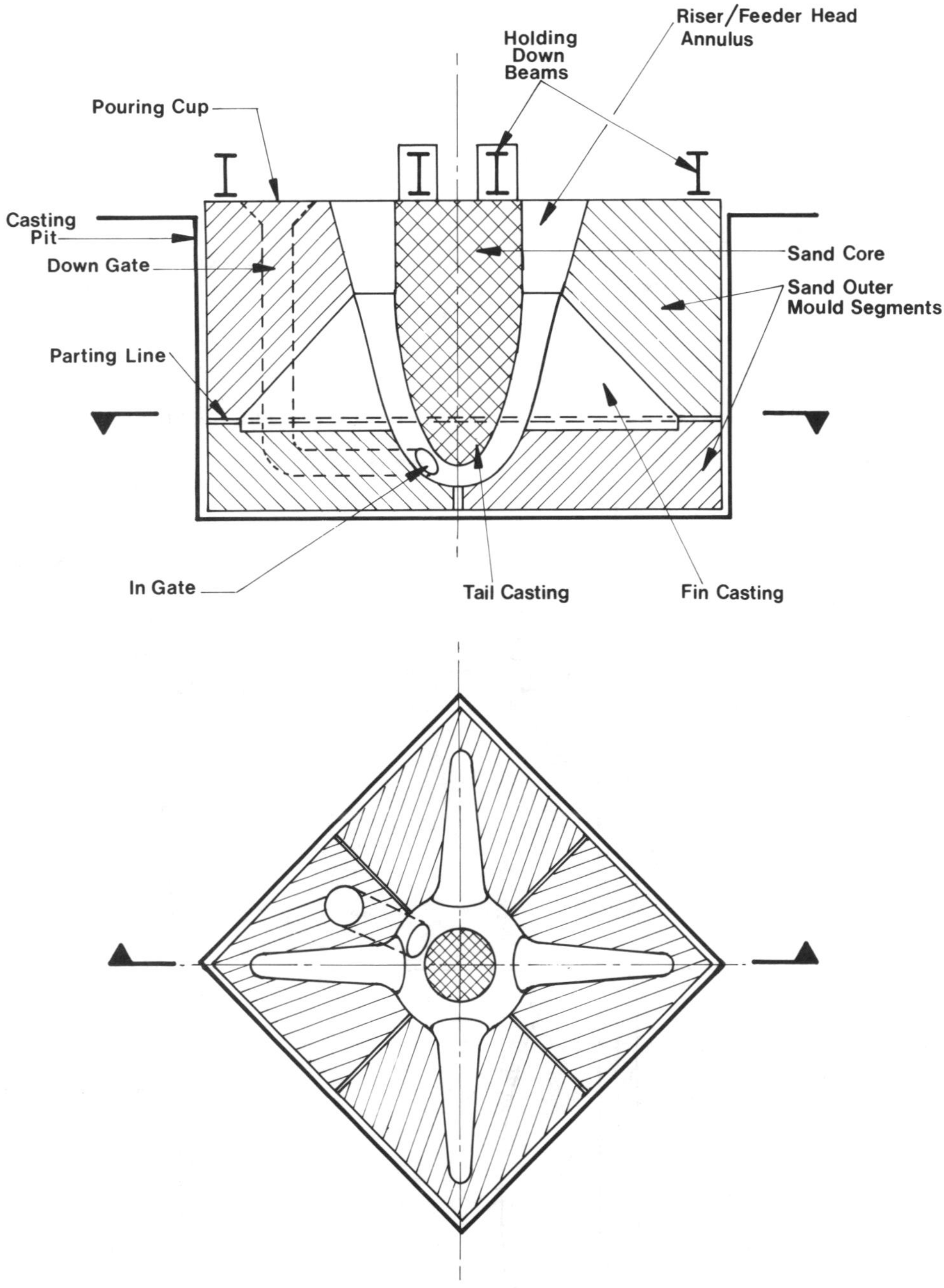

TAIL MANUFACTURE : SAND CASTING **figure 26**

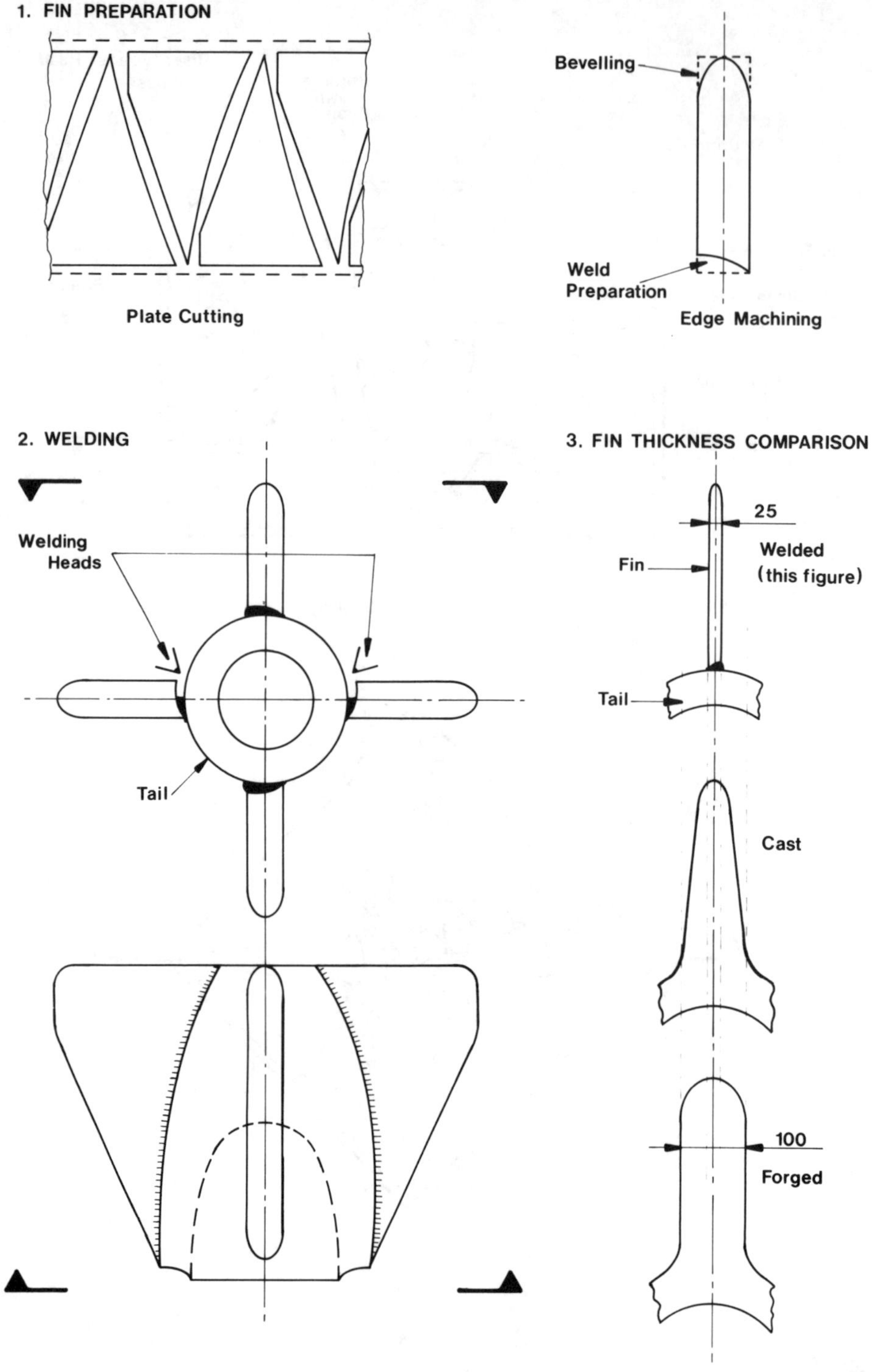

TAIL MANUFACTURE: WELDED FINS

figure **27**

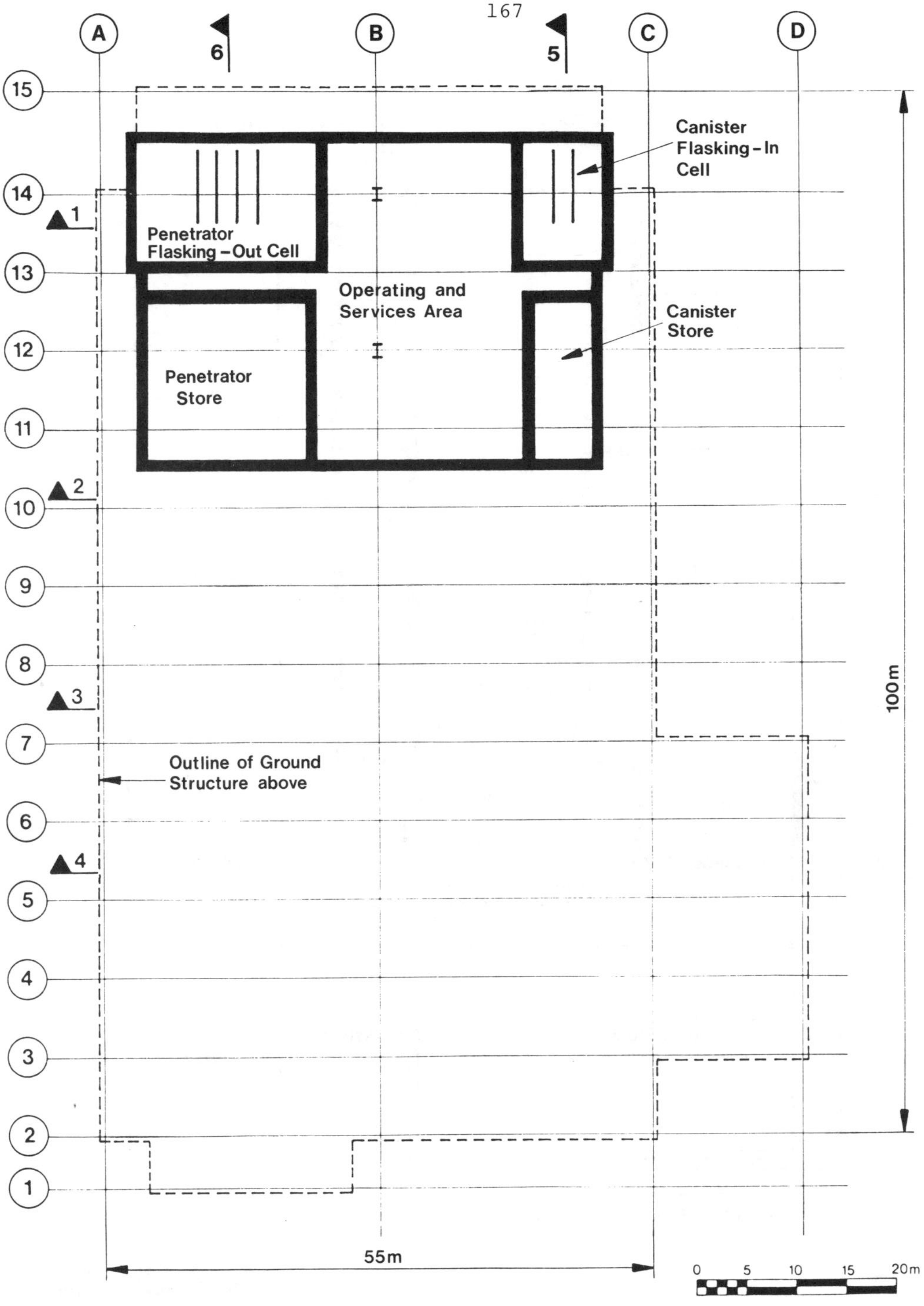

ASSEMBLY PLANT : BASEMENT FLOOR PLAN

figure 28

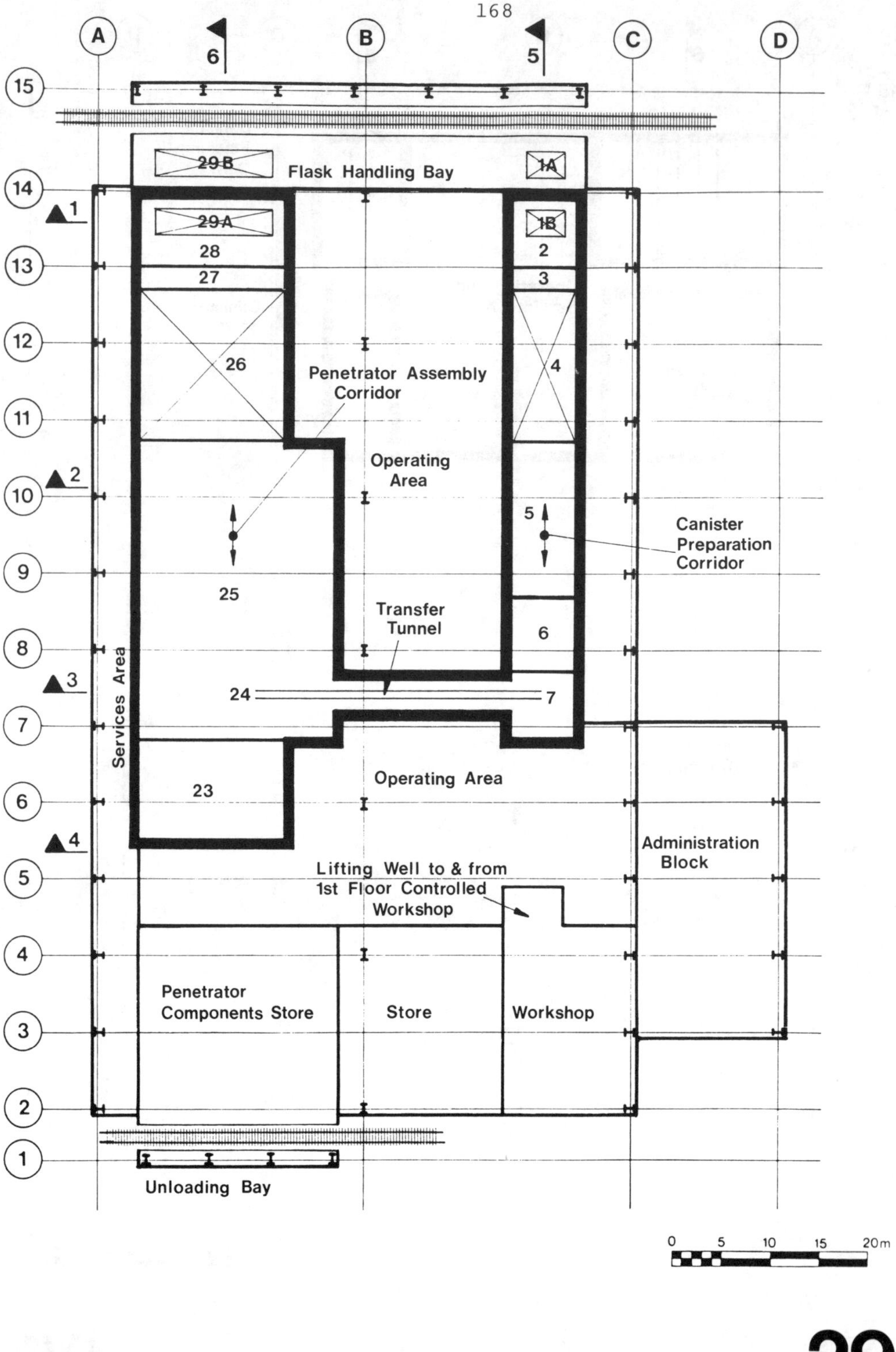

ASSEMBLY PLANT : GROUND FLOOR PLAN

figure 29

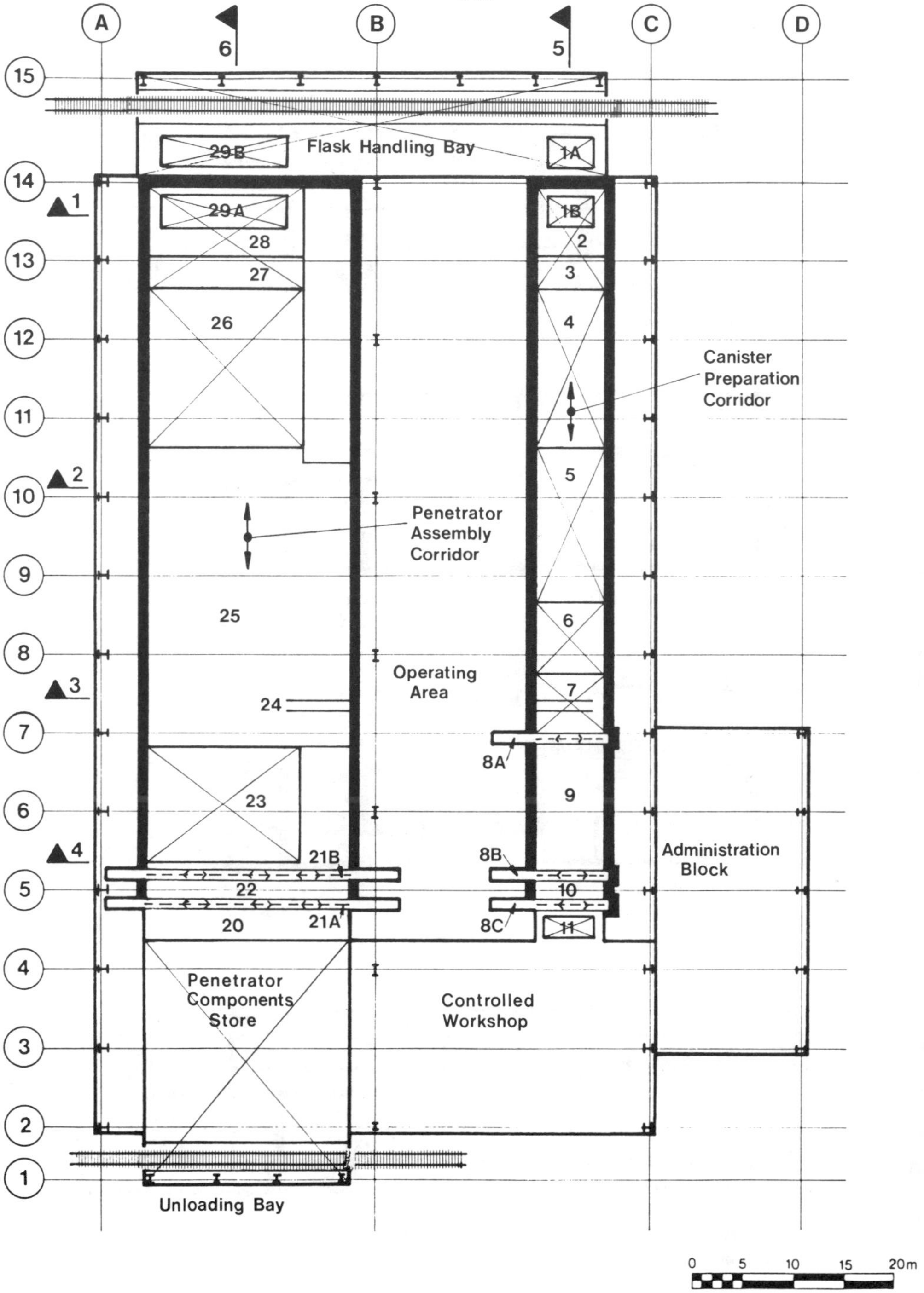

ASSEMBLY PLANT: FIRST FLOOR PLAN

figure **30**

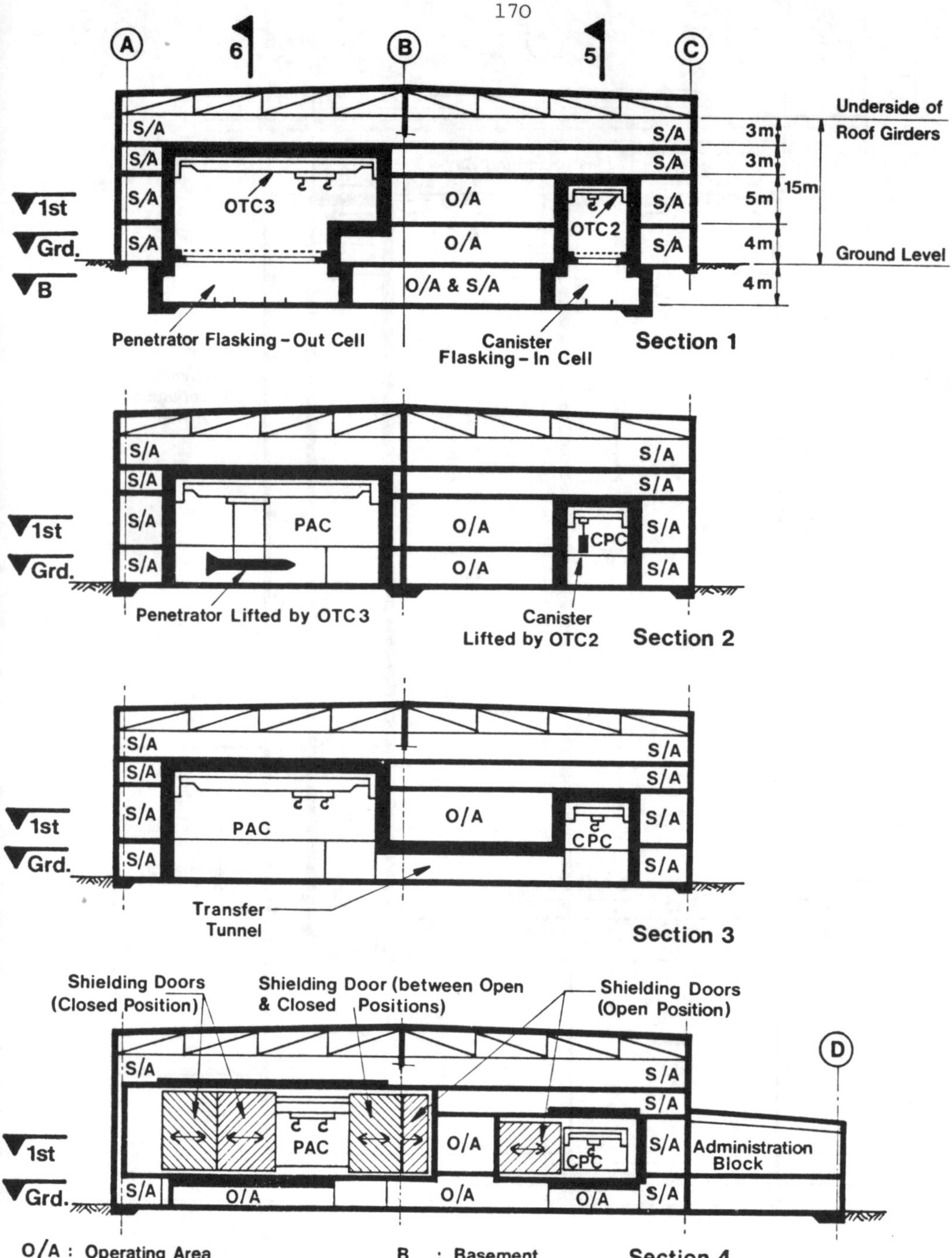

O/A : Operating Area
S/A : Services Area
OTC : Overhead Travelling Crane
CPC : Canister Preparation Corridor
PAC : Penetrator Assembly Corridor

B : Basement
Grd. : Ground Floor
1st : First Floor

ASSEMBLY PLANT: SECTIONS 1 - 4

figure 31

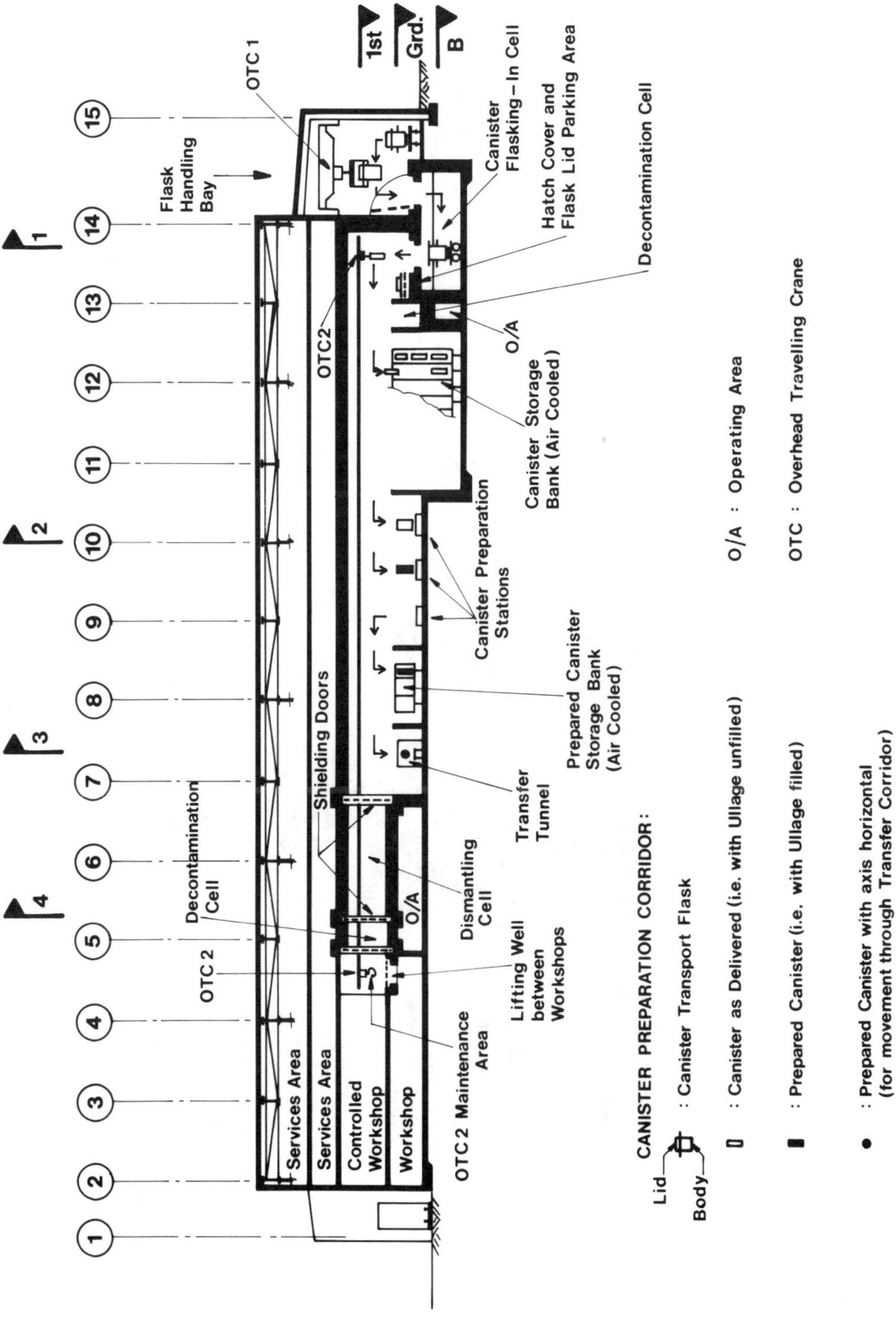

ASSEMBLY PLANT : SECTION 5

figure 32

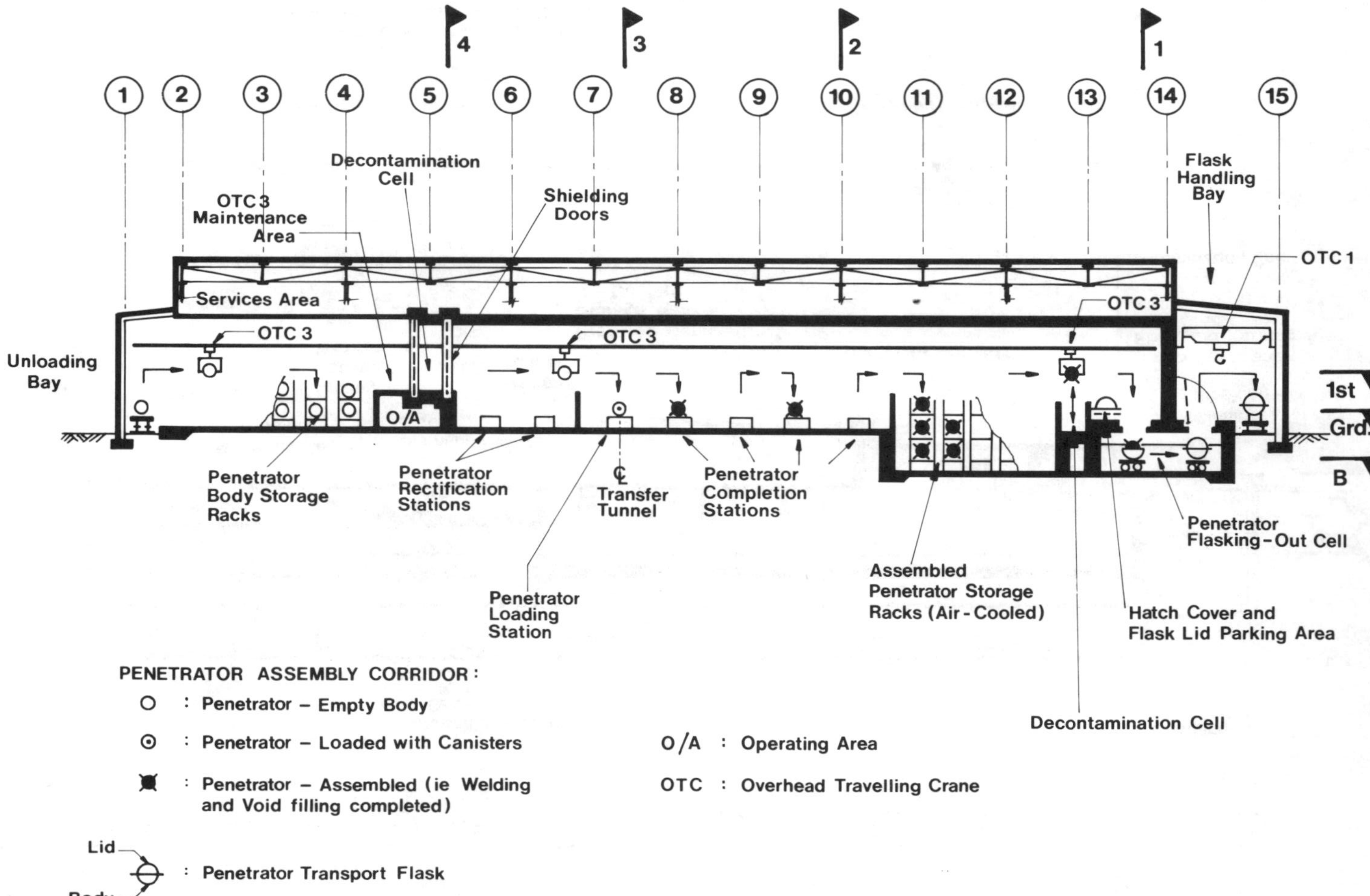

ASSEMBLY PLANT : SECTION 6

figure 33

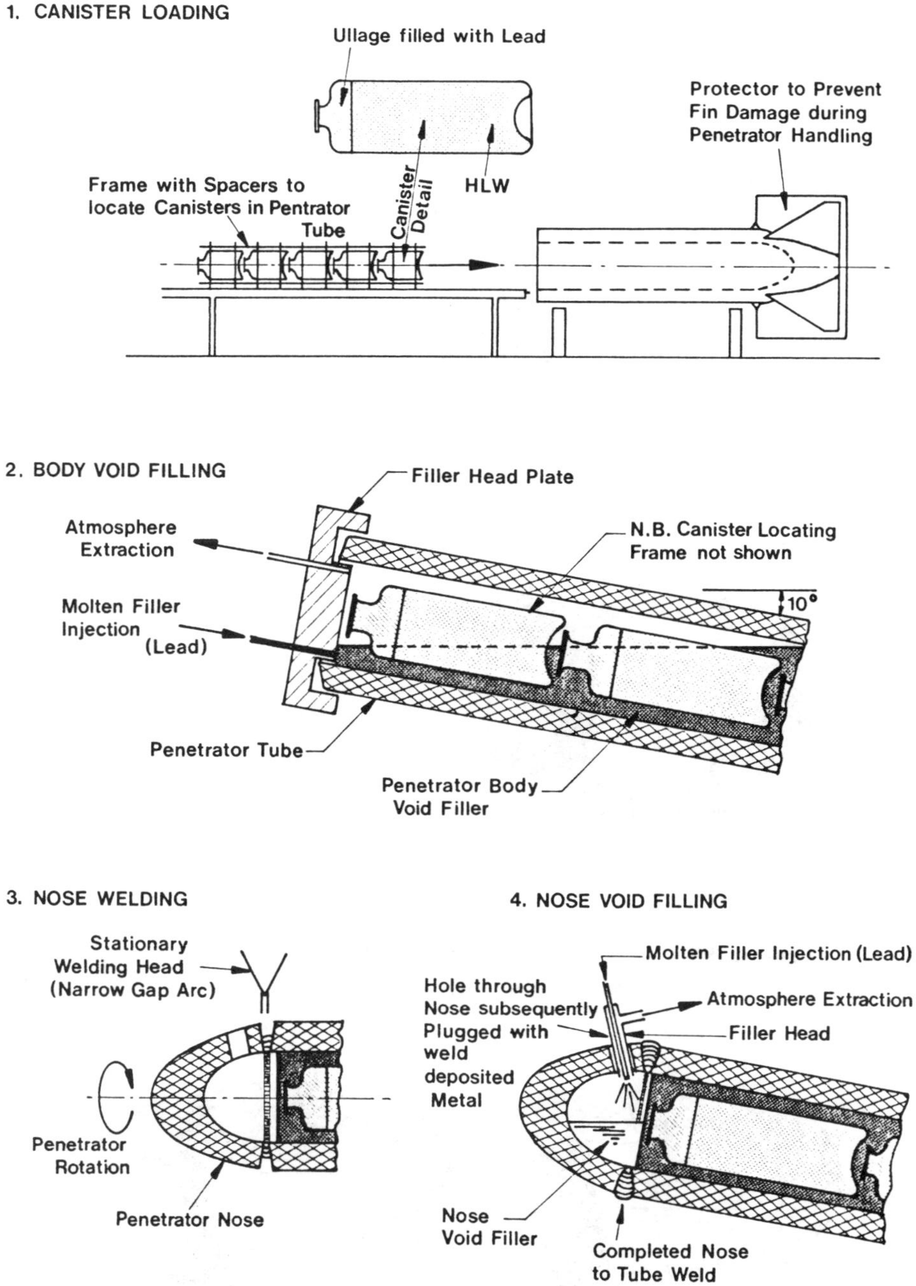

PENETRATOR ASSEMBLY

figure 34

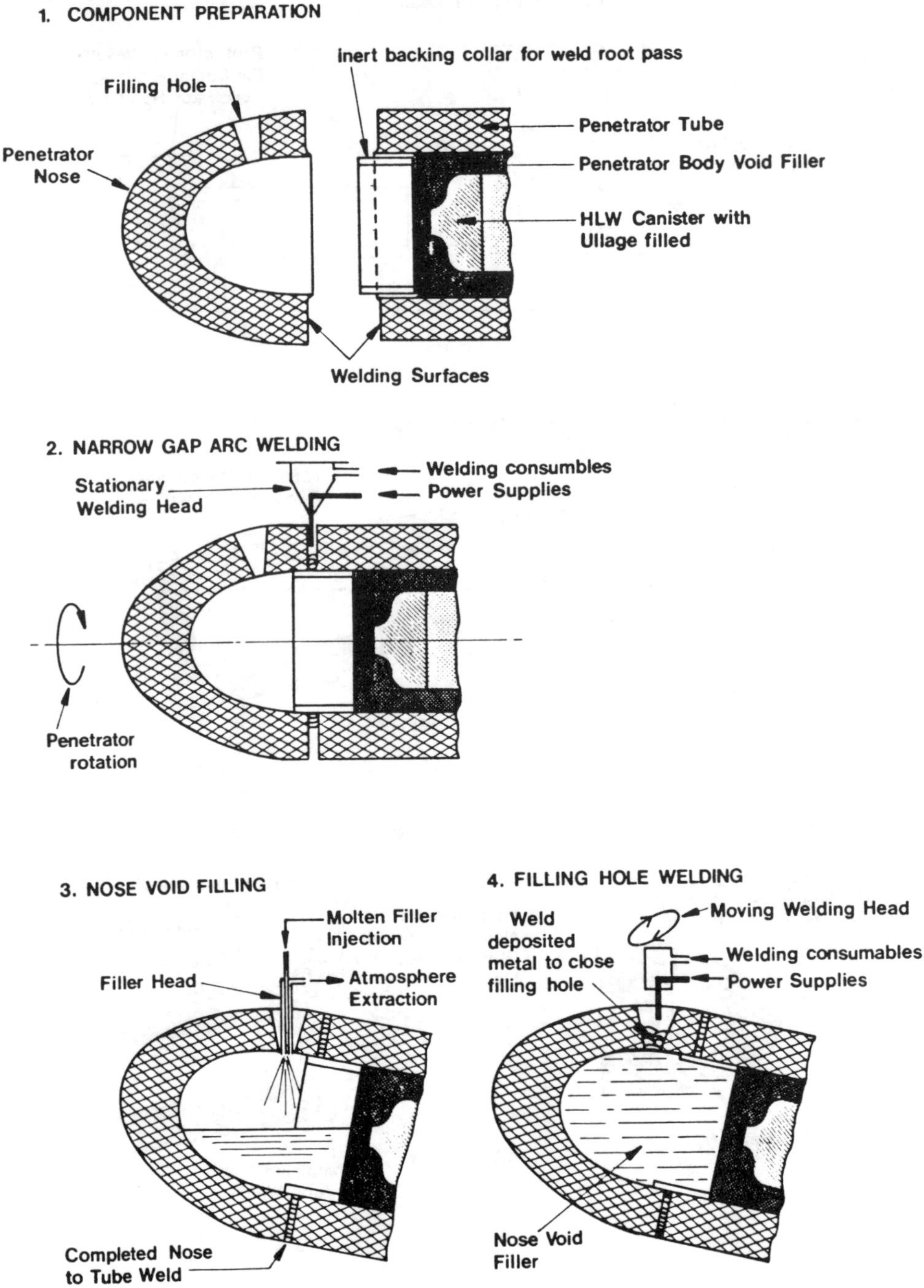

PENETRATOR COMPLETION: OPTION A

figure 35

1. COMPONENT PREPARATION

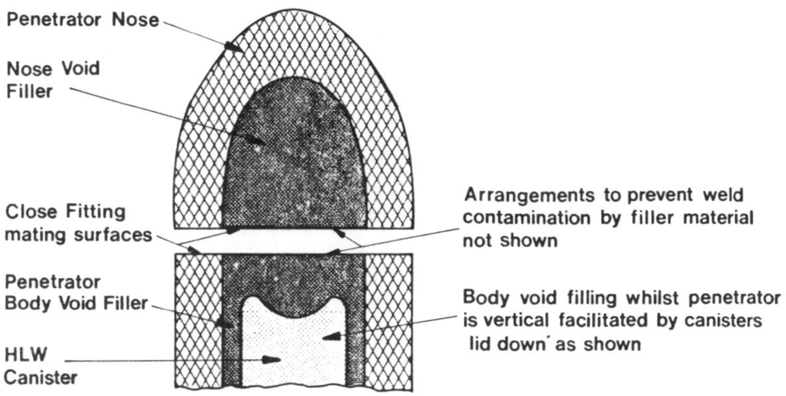

2. ELECTRON BEAM WELDING

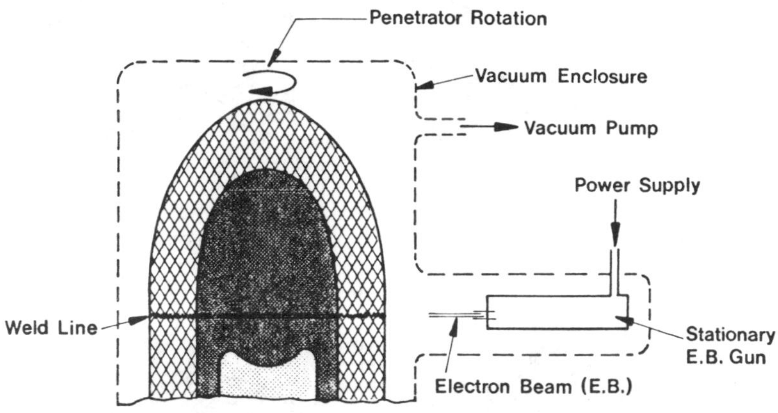

PENETRATOR COMPLETION: OPTION B

figure 36

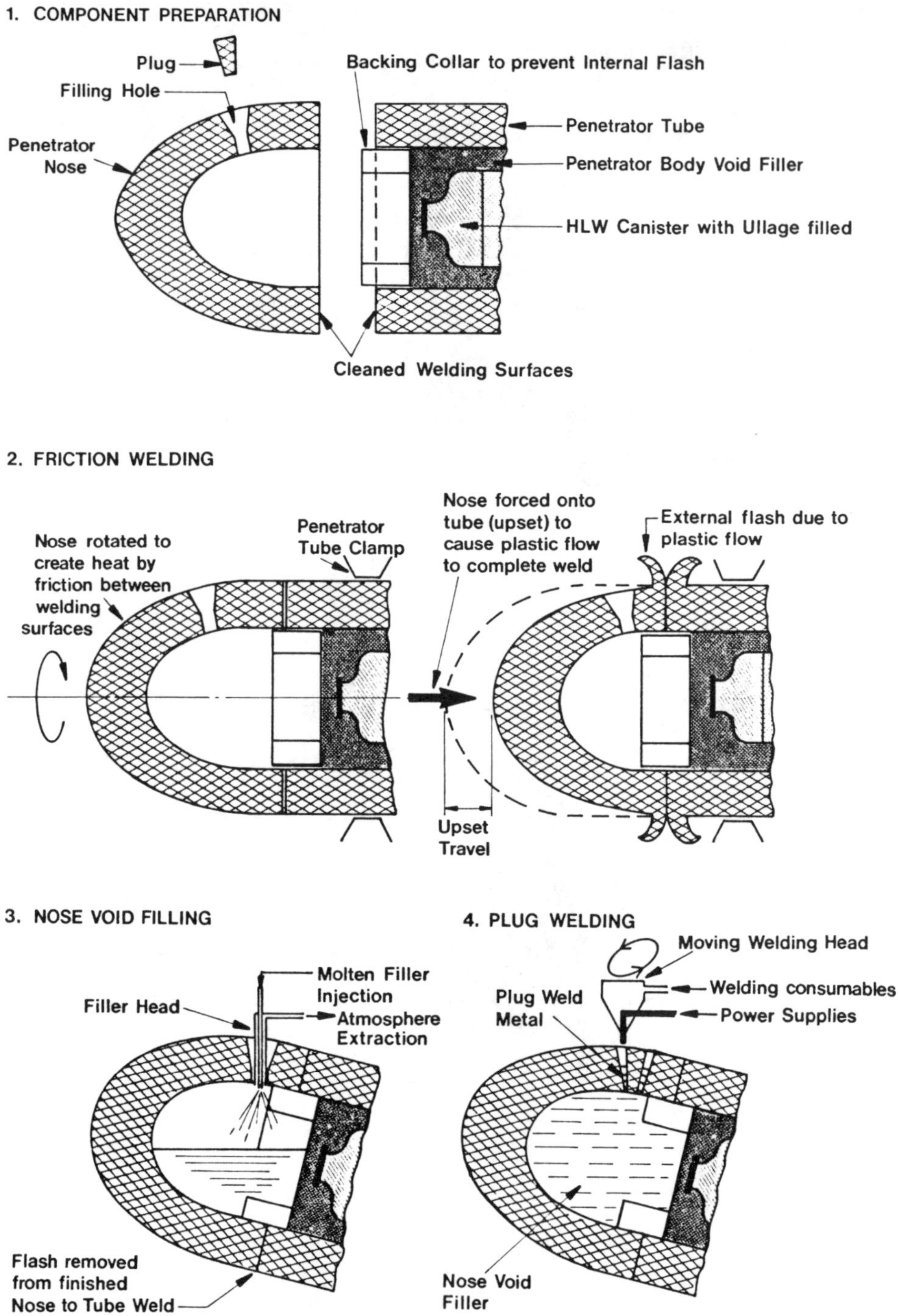

PENETRATOR COMPLETION:OPTION C

figure **37**

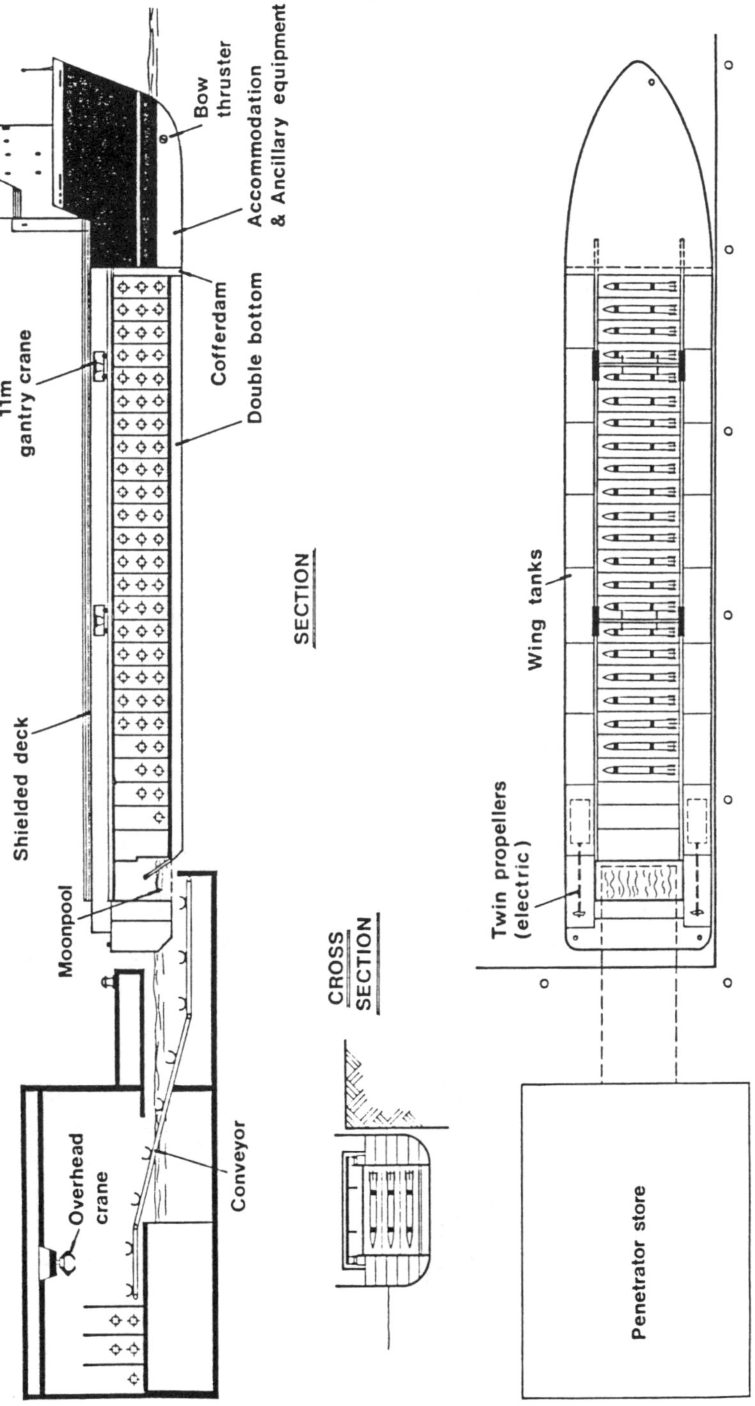

DISPOSAL SHIP OPTION A

figure 38

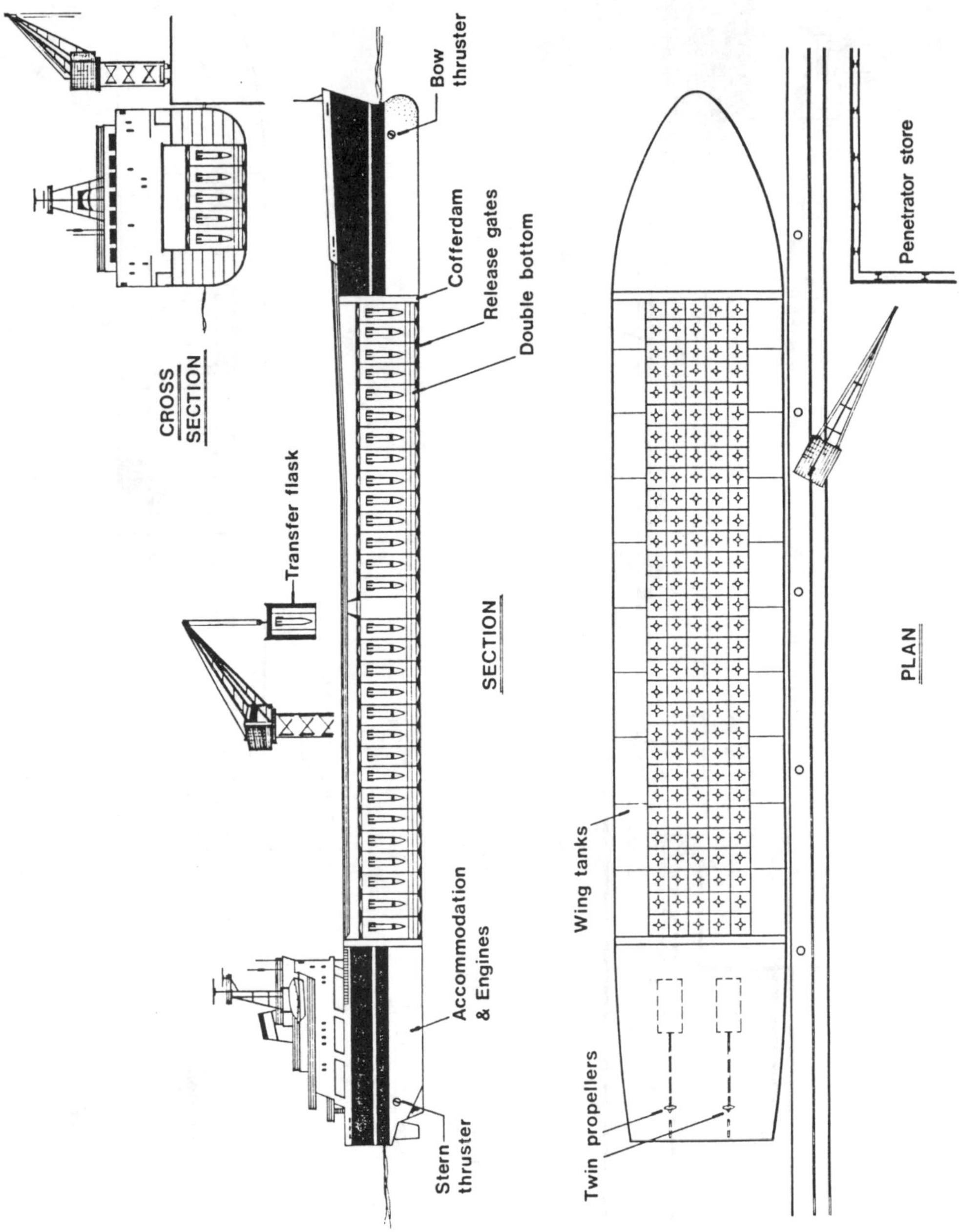

DISPOSAL SHIP OPTION B

figure 39

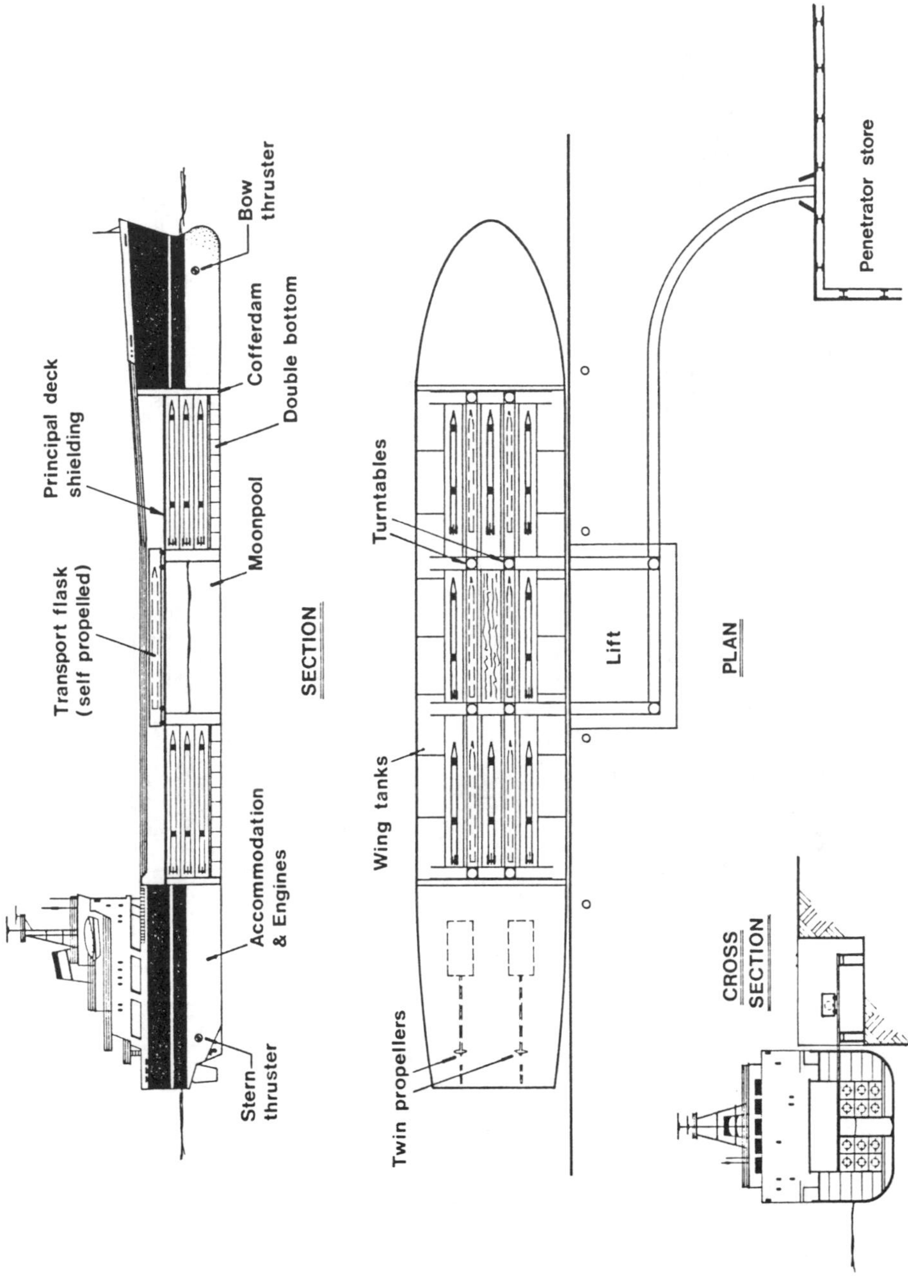

DISPOSAL SHIP OPTION C

figure 40

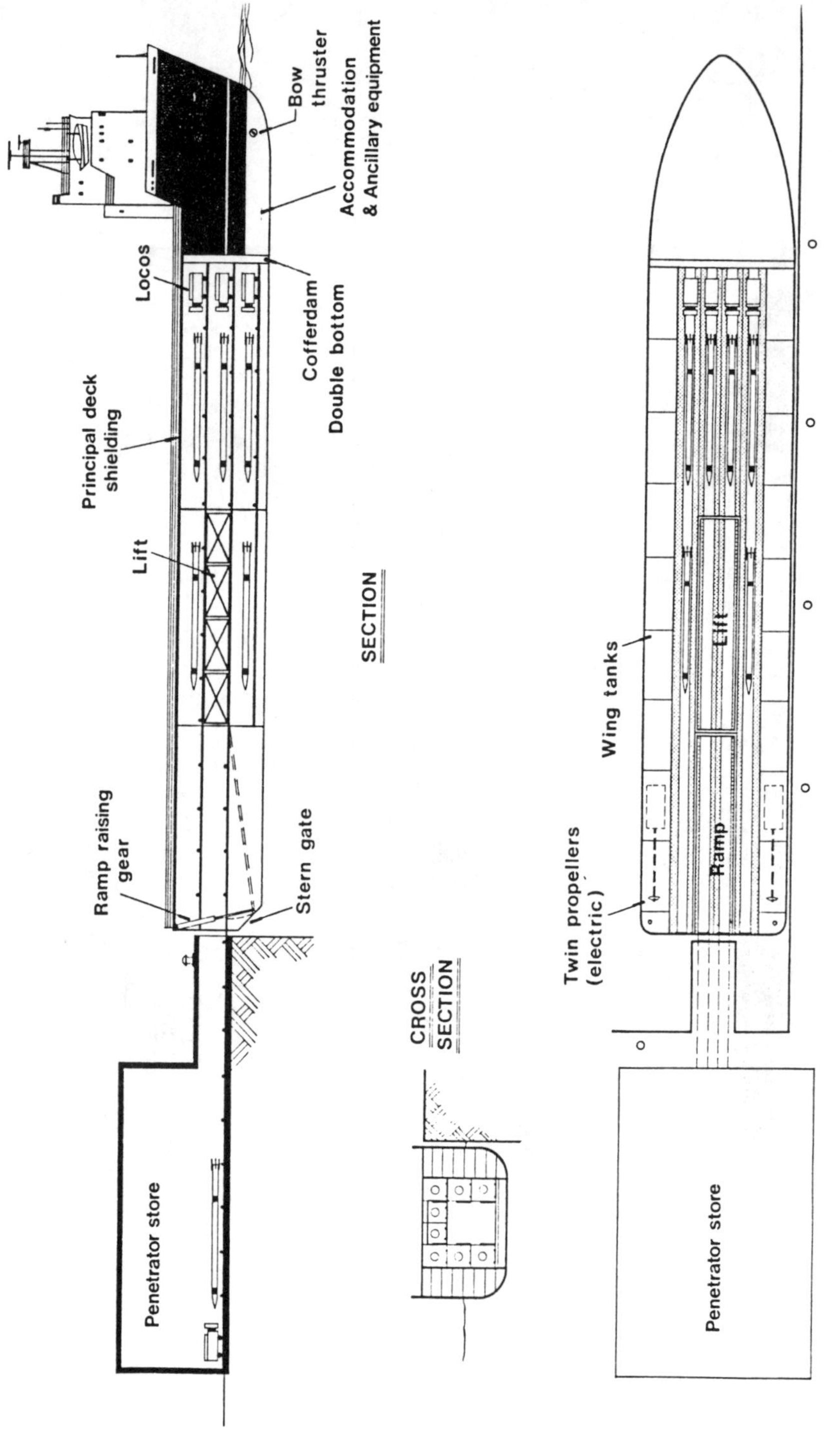

DISPOSAL SHIP OPTION D

figure 41

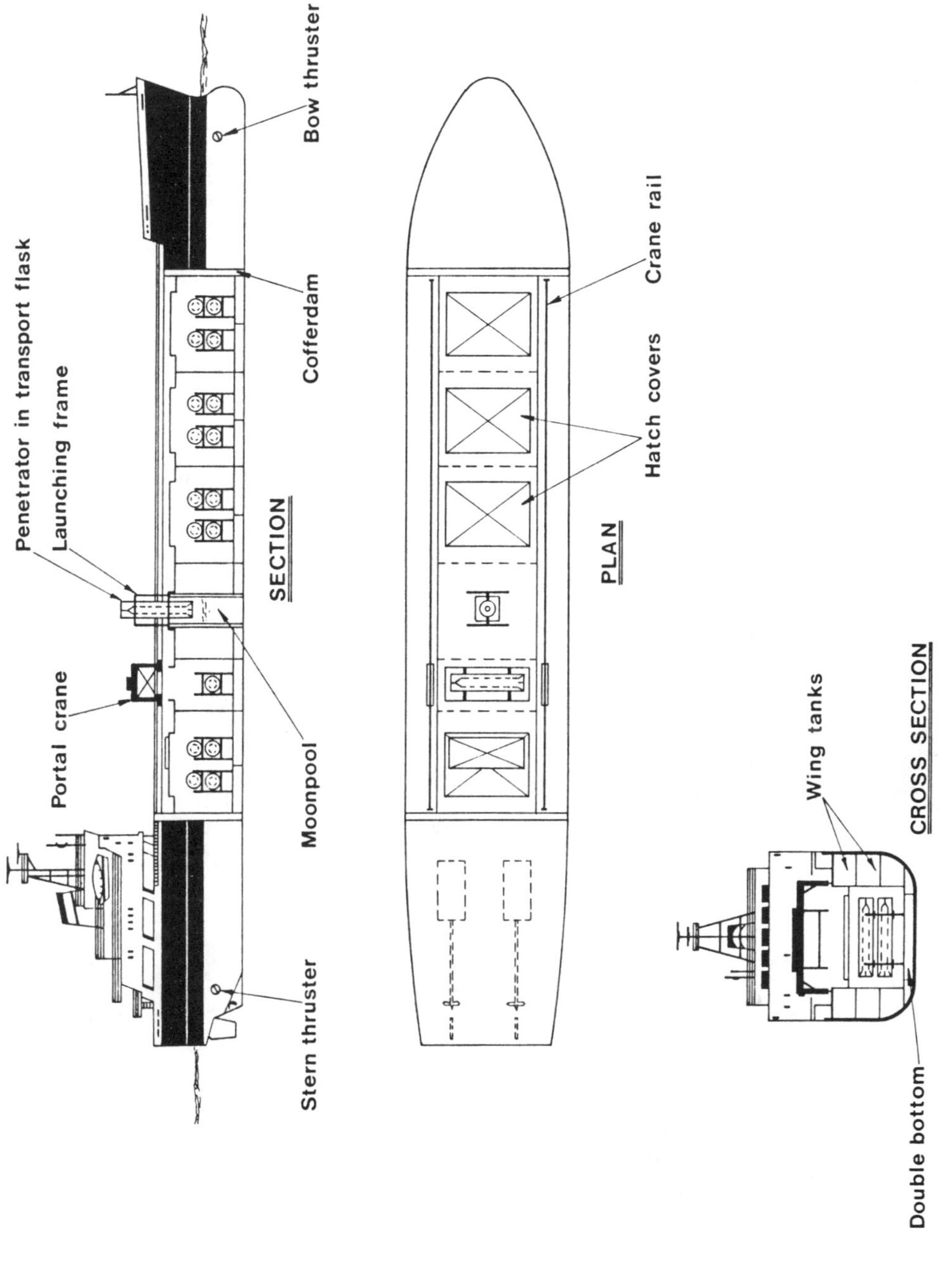

DISPOSAL SHIP OPTION E

figure 42

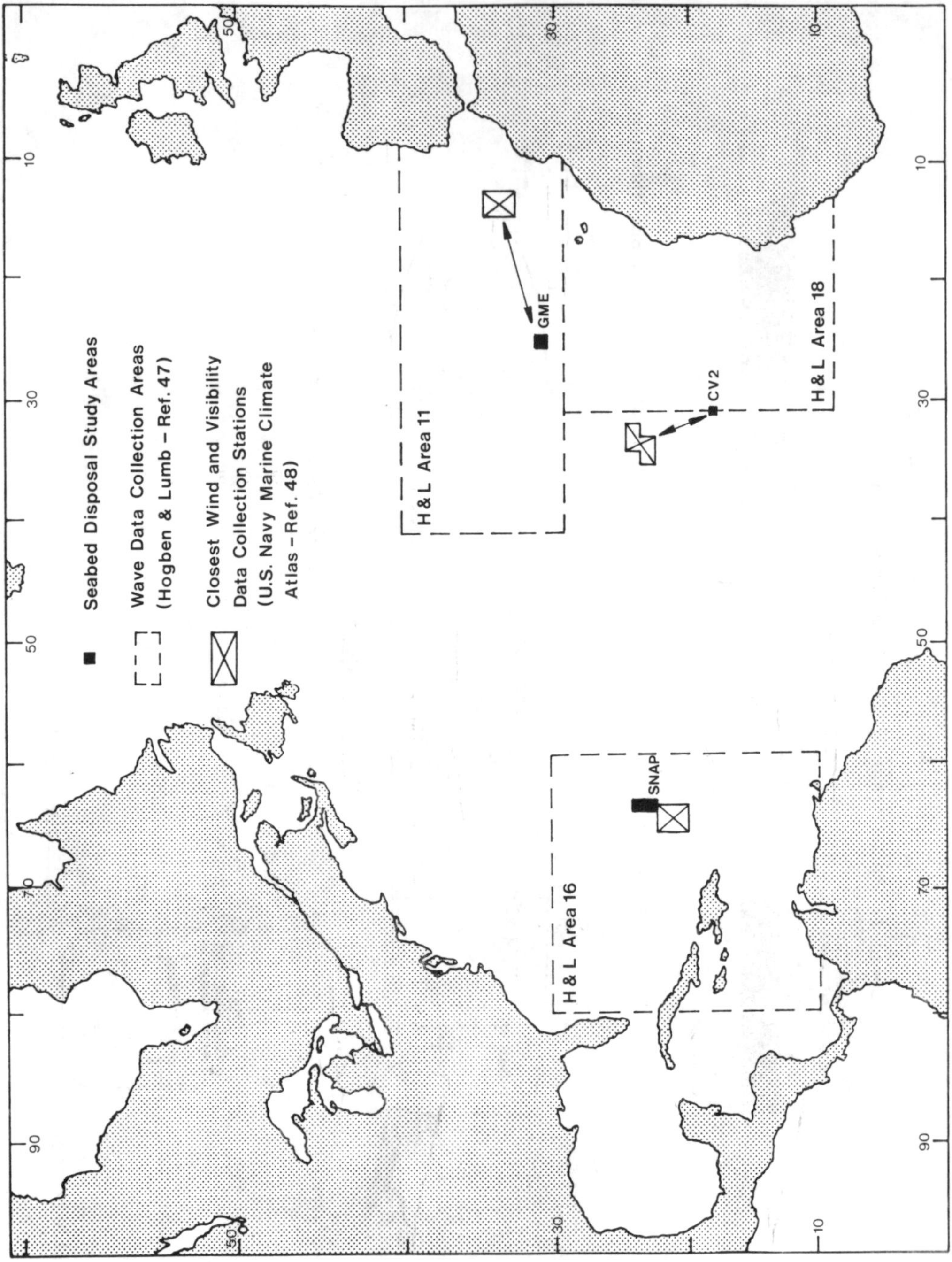

NORTH ATLANTIC WEATHER DATA SOURCES figure 43

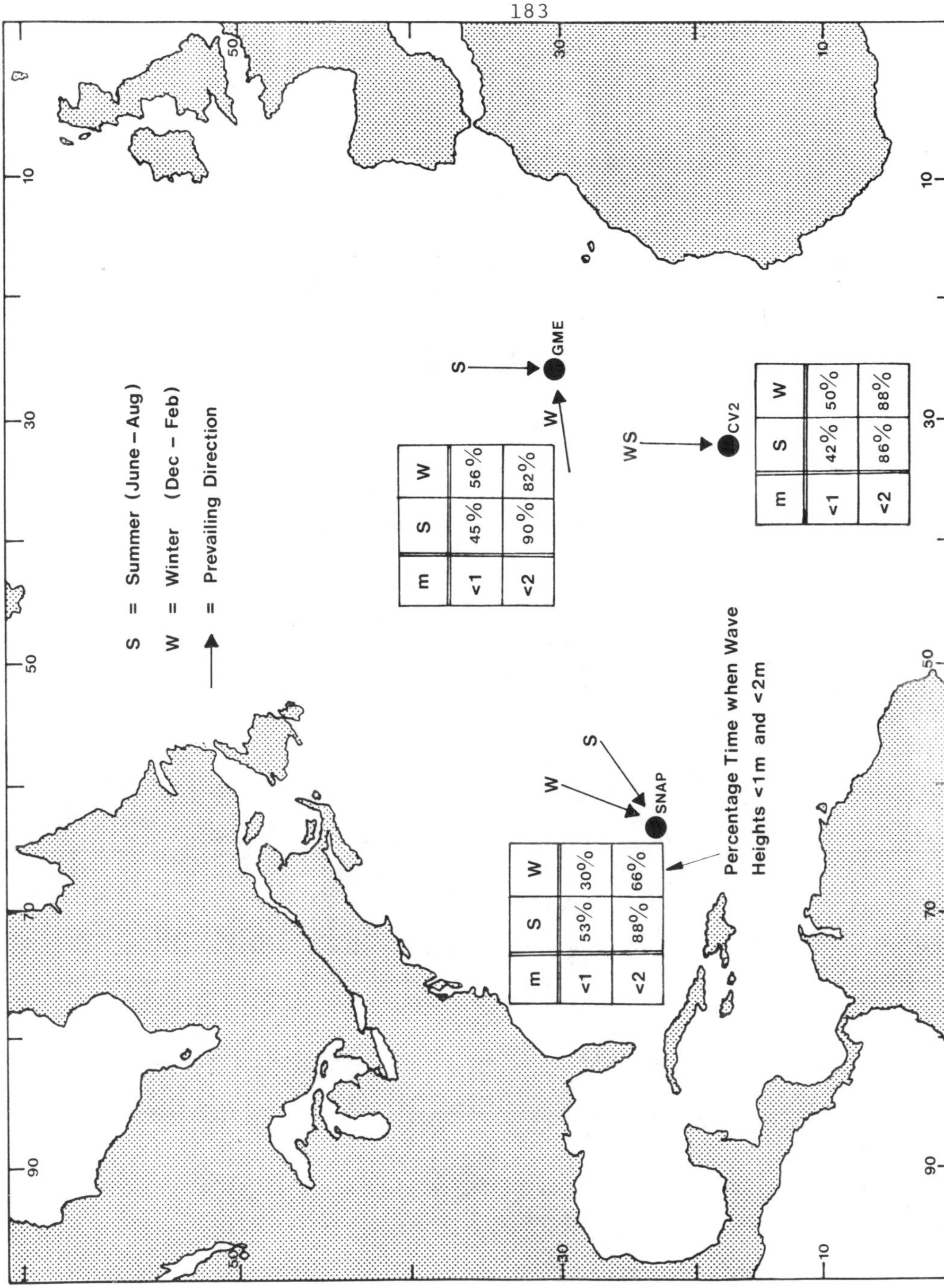

NORTH ATLANTIC WAVES

figure 44

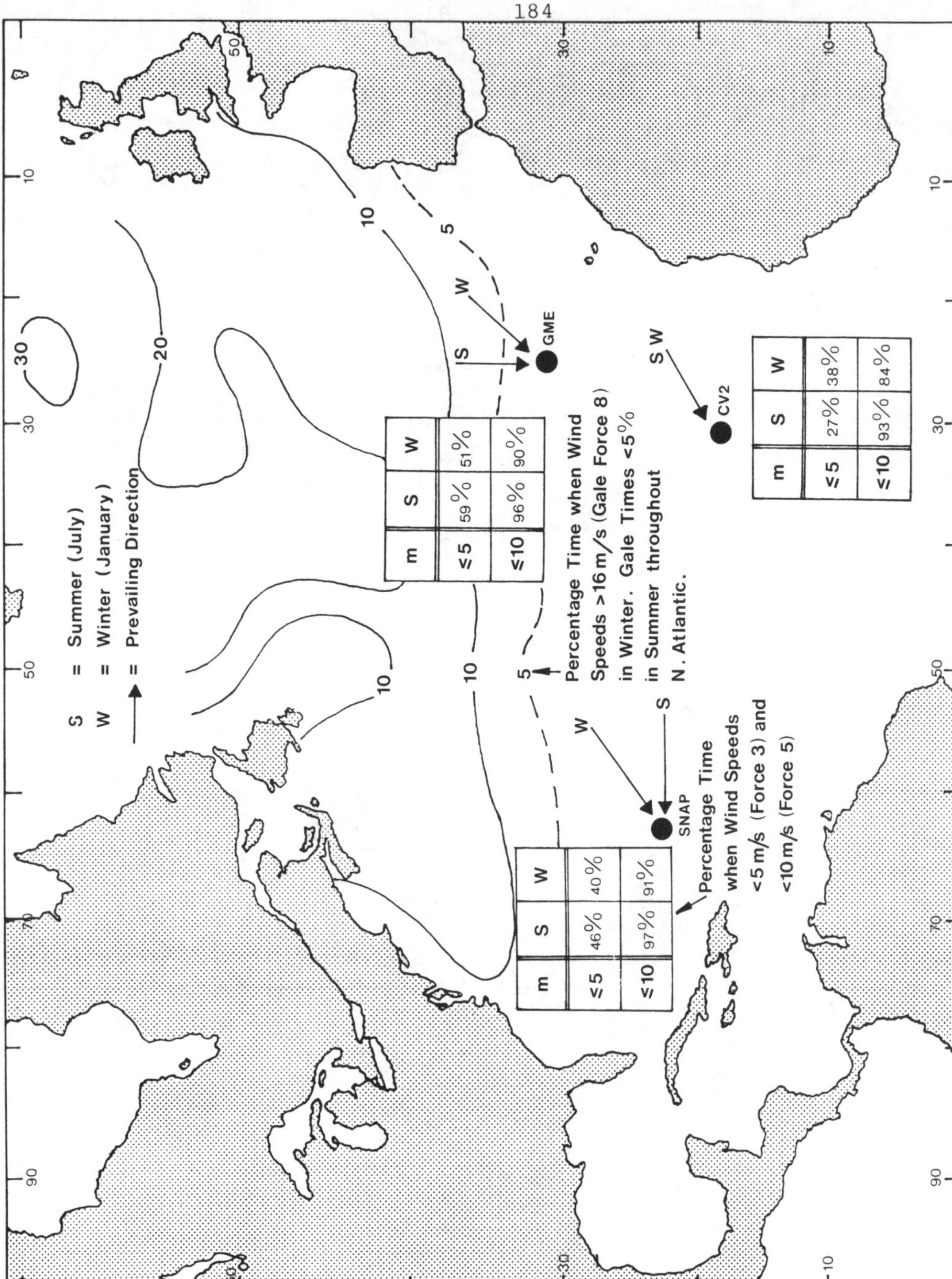

NORTH ATLANTIC WINDS

figure 45

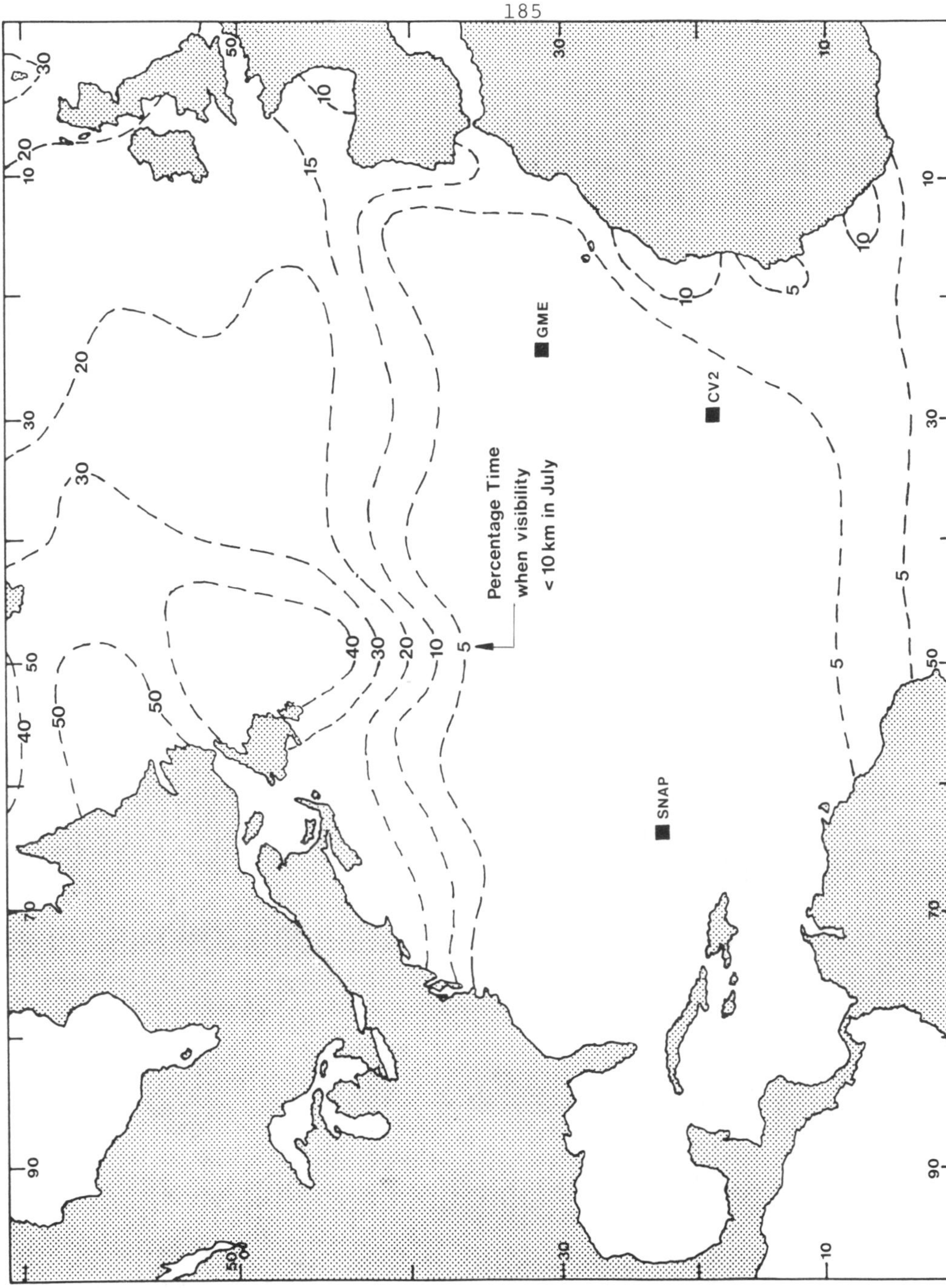

NORTH ATLANTIC SUMMER VISIBILITY

figure **46**

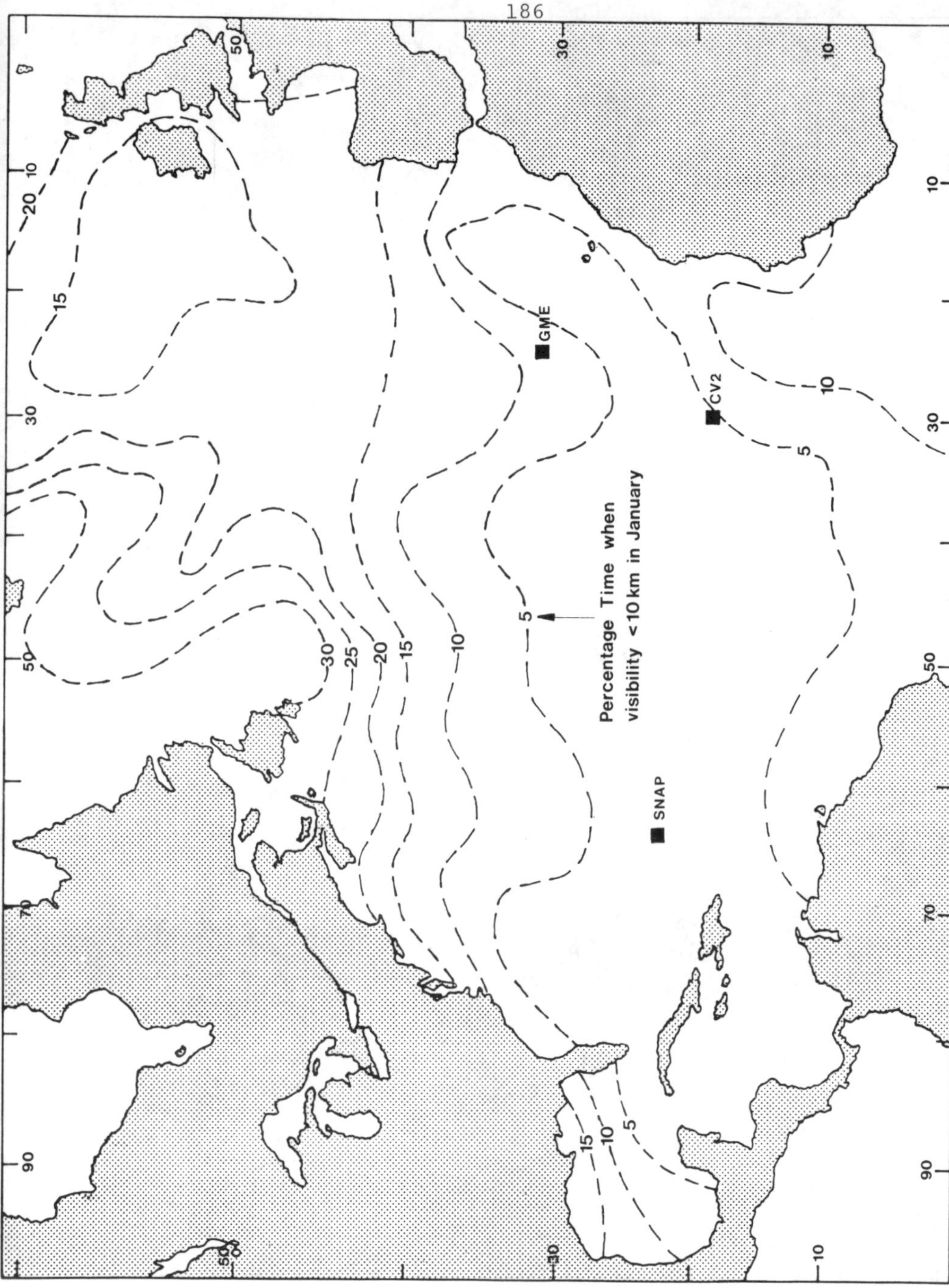

NORTH ATLANTIC WINTER VISIBILITY

figure 47

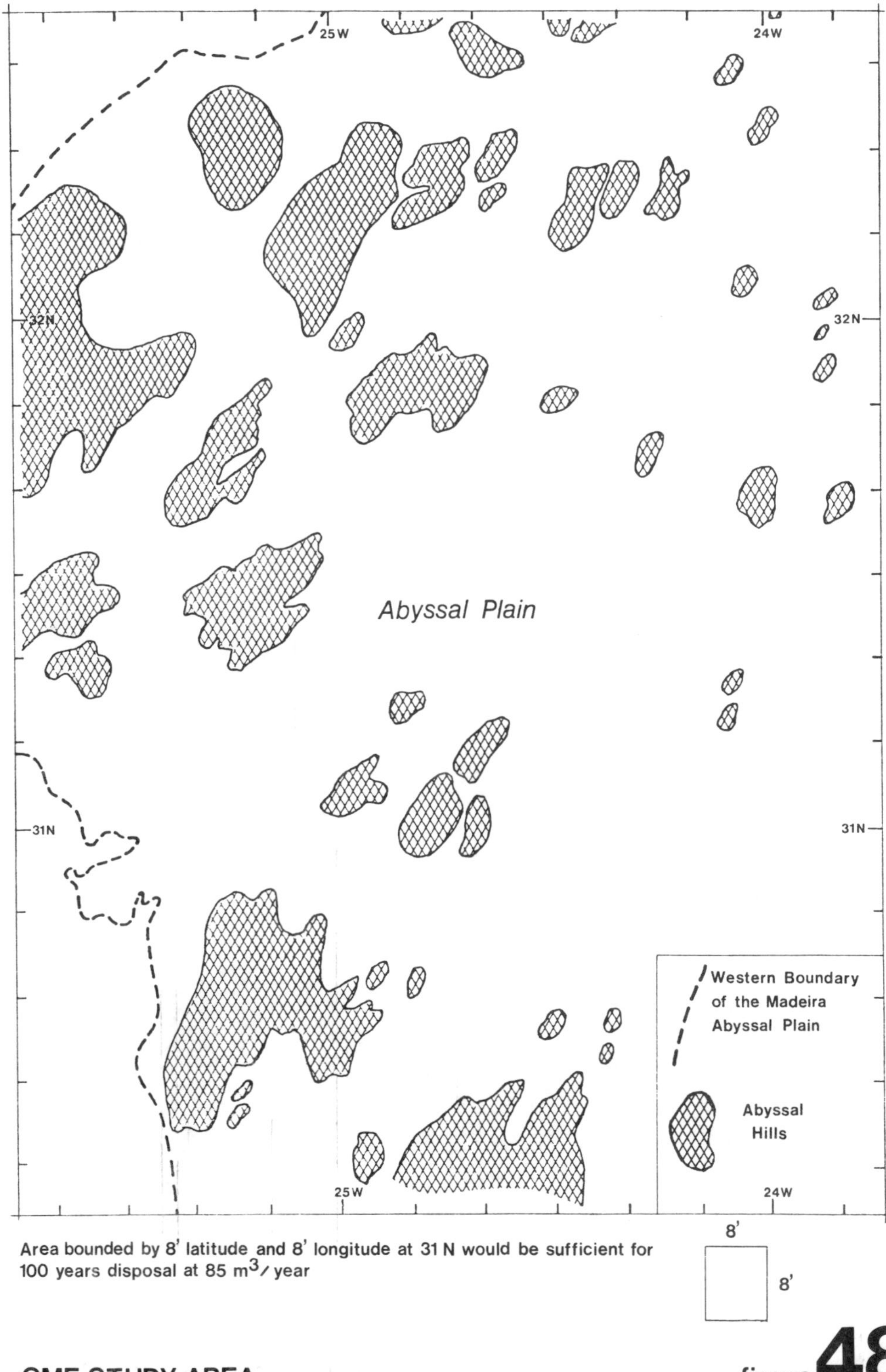

Area bounded by 8' latitude and 8' longitude at 31 N would be sufficient for 100 years disposal at 85 m^3/ year

GME STUDY AREA

figure 48

CONTENTS OF APPENDICES

APPENDIX A

BASIC DATA

APPENDIX A

BASIC DATA

A.1 Introduction

This Appendix collects and presents in concise form the various data and assumptions which formed the basis for the penetrator engineering study.

HLW management systems are in a relatively early state of development and very few common parameters have as yet found universal acceptance. For this reason only single typical values of data are given.

A.2 HLW Characteristics

Unless otherwise noted, the following data have been extracted from "Specifications of Vitrified Residues Produced from Reprocessing at UP2 or UP3-A La Hague Plants" published by COGEMA in September 1982 for the vitrification of HLW into a borosilicate glass form.

A.2.1 Nominal spent fuel data

a) 3.5% U_{235} enriched PWR type 17x17 fuel

b) 33000 MWd/tU average burn-up

c) Reprocessed 3 years after discharge from reactor.

A.2.2 Nominal HLW vitrification data

a) HLW vitrified 1 year after fuel reprocessing, that is 4 years after fuel discharged from reactor

b) HLW canister details given in Figure 2.

c) HLW heat generation decay given in Figure 3.

d) Sum of oxides of fission products and actinides comprises 12% by weight of vitrified HLW.

e) $1.93m^3$ HLW produced for 1 GW(e)y power generation based on performance estimate by the Commission of the European Communities for a 1GW(e) generic plant.

f) HLW density = $2.7t/m^3$.

g) Minimum transformation temperature = 510°C.

h) Thermal conductivity = 1.165 W/m/k at 150°C.

A.3 HLW Disposal Programmes

Representative HLW vitrification rates for consideration in engineering studies were provided in a private communication by the Commission of the European Communities based on information furnished by member states. The two CEC rates, high and low, correspond to the maximum and minimum nuclear power installation growth plans to the year 2000 of all the Community members.

A third disposal rate, nominally set at half the CEC-low figure, has been chosen as a conjectural HLW disposal programme to allow for the possibility that not all of the Community's waste would be disposed of to the ocean. The third rate also corresponds to the full potential UK disposal rate at the year 2000.

A.3.1 Disposal rates

a) CEC - high rate : 400m^3 HLW per year

b) CEC - low rate : 170m^3 HLW per year

c) CEC - 50% low rate)
UK - full rate) : 85m^3 HLW per year

A.3.2 Pre-disposal storage

a) A 50 year nominal interim storage period for the heat generation and radioactivity of HLW to decay to suitable levels for disposal has been assumed. The earliest commissioning date for any ocean disposal programmes would, therefore, be the middle of the next century.

b) The consequences of disposal for shorter or longer interim storage periods are discussed where appropriate.

A.4 Thermal Modelling

The following data have been used for predicting temperatures around penetrators emplaced in the seabed sediments.

a) Ambient temperature at seafloor = 4°C

b) Temperature increase with depth into seabed = 54°C/km.

c) Seabed thermal conductivity = 0.8 W/m/K.

d) Seabed thermal capacity = 3780 J/kg/K.

A.5 Seabed Disposal Study Areas

The seabed sediments in several areas of the Atlantic and Pacific oceans are being investigated to gather information for general HLW disposal research work. The three North Atlantic areas described below have been considered in this study to give typical sailing distance and weather data for notional HLW disposal sites. There is no suggestion at this stage that any of them would be for used in an HLW ocean disposal programme.

A.5.1 Location

Study Area	Location	Ocean Depth
Great Meteor East (GME)	31.2° - 31.8° N 24.2° - 25.0° W	5400m
Cape Verde Rise (CV2)	19.1° - 19.5° N 29.6° - 30.0° W	4700m
Southern Nares Abyssal Plain (SNAP)	22.5° - 23.5° N 62.5° - 65.0° W	5800- 5900m

Sailing distances to the study areas are given in Figure 1.

A.5.2 Weather

Sources of weather data are given in Figure 43 while wave, wind and visibility data for the North Atlantic are given on Figures 44 to 47.

APPENDIX B

HYDRODYNAMIC AND PENETRATION MODEL DEVELOPMENTS

APPENDIX B

HYDRODYNAMIC AND PENETRATION MODEL DEVELOPMENTS

B.1 Introduction

Certain refinements have been introduced to the previously developed hydrodynamic and embedment models (Ref. 7) for the purposes of this study.

B.2 Hydrodynamic Model

The model treats all the various drag effects as being divided into two components. Namely, a skin friction component dependent on the wetted surface area of the penetrator and a residual component, dependent on the frontal area, which covers all other effects including pressure drag and drag from interactions between body and tail fins.

Discussion with the Admiralty Marine Technology Establishment (AMTE) together with further literature study, suggested that the previous value of the residual drag coefficient was over-conservative for the slender penetrator vehicle. Use of Figure 2 from Granville's Report (Ref. 49) leads to a choice of residual drag coefficient $C_r = 0.66$, to suitably cover a range of penetrator geometries. The author states that this figure, although taken from empirical data for torpedoes, represents the behaviour of average operational underwater bodies and may be used to estimate the value of the coefficient.

The skin friction coefficient is, as previously, obtained from a flat plate theory equation such as Schoenherr's (Ref. 50) given below.

$$\frac{1}{C_s} = 3.46 \log_{10} (Re) - 5.6$$

Re is the Reynolds number of the penetrator. AMTE indicated that Re should be calculated using penetrator length rather than diameter, as this is more appropriate for slender bodies.

Therefore,

$$Re = \frac{U_o L}{\nu}$$

Where U_o = terminal velocity
L = penetrator total length
ν = kinematic viscosity of sea water

A value of skin friction coefficient $C_s = 0.002$ has been adopted, again to cover a range of penetrator geometries.

Use of these coefficients causes an increase of approximately 20% in the terminal velocity compared with the velocity predicted using the coefficients of the previous OAP report (Ref. 7) which were $C_s = 0.0025$ and C_r (previously written as C_D) = 0.15.

B.3 Penetration Model

Two revisions have been made to the penetration model employed previously (Ref. 7). The first of these relates to the undrained shear strength profile of the sediment, and the second relates to consideration of hydrodynamic resistance during the partial embedment stage of penetration.

The undrained shear strength profile employed in preliminary studies was:

$$c_u = 1.5z \text{ kN/m}^2$$

Where z = depth below seabed in metres.

This profile does not take into account the increase in density of the sediment with depth, and attributing zero strength to the sediment at the seabed is possibly unreasonable, particularly when consideration is given to sediment inertia effects at shallow embedment depths.

With these considerations in mind, a 'tri-linear' profile was developed. However, for the purposes of the feasibility study a less complex linear profile has been developed as a best fit to the 'tri-linear' profile. The adopted profile is

$$c_u = 5 + 1.5z \text{ kN/m}^2$$

Use of this profile causes a decrease in predicted embedment depth of between 5 and 15% compared with the previous model.

The penetration model has been revised to include hydrodynamic resistance acting on the penetrator during the stages of partial embedment. The model now includes both skin friction drag acting on the penetrator surface area remaining above the mudline, as well as pressure drag arising from flow separation at the penetrator tail.

Pressure drag is assumed to comprise 90% of the total residual drag.

Inclusion of these terms causes a nominal 1 to 2% reduction in the estimated embedment depth.

B.4 Fin Design

A recent computer modelling study of penetrator hydrodynamics (Ref. 21) has shown that generally a set of fin misalignments, causes a spiral trajectory, and that a fin twist of just one degree would be sufficient in still water to control the horizontal displacement of the reference penetrator at the seabed to less than 15 metres from its launch point.

In the same study the maximum load on a reference penetrator tail fin was predicted to be approximately 1 tonne applied close to the leading edge immediately after entry into the sediment.

APPENDIX C

PENETRATOR SPECIFICATION DETAILS

APPENDIX C

PENETRATOR SPECIFICATION DETAILS

C.1 Steel Chemical Composition

A guide chemical composition has been assumed as follows:

0.15 wt. % max. carbon (C)
0.40 wt. % max. silicon (Si)
1.50 wt. % max. manganese (Mn)
0.02 wt. % max. phosphorous (P)
0.01 wt. % max. sulphur (S)
1.00 wt. % max. sum total of residual elements
0.40 wt. % max. any individual residual element
Balance iron (Fe)

Notes on the various aspects of the steel specification are as follows:-

a) killing of steel refers to the elimination of iron oxide, that is deoxidation, during the refining stage

b) electric quality steel is steel made in an induction or lined electric arc furnace and is distinguished by its high purity and low sulphur and phosphorus content. Electrically melted steel describes steel which has been refined in an electric furnace to less strict requirements than for electric quality, being equivalent instead to an open hearth steel

c) double slag refining is the standard electric arc furnace process for high quality steelmaking

d) secondary steelmaking processes are used for further refinement of steels from conventional furnaces to produce specially high standard material grades

e) vacuum degassing is a procedure which further improves the microstructure of high quality steels

f) chemical composition of the steel is expressed in terms of the maximum permitted percentages by weight of alloying elements. Precise alloy compositions are normally determined by the steelmaker to achieve the mechanical properties specified for the steel but also vary slightly between furnace melts or batches

g) the carbon content is selected mainly to ensure a readily weldable steel. A 0.15% C steel is considered to be a low carbon grade

h) a silicon content is included from the point of view of steel killing

i) a manganese content is selected in combination with the carbon content for the achievement of mechanical properties, principally yield and tensile strengths, normally specified for general engineering steels

j) phosphorus and sulphur contents are set at very low levels to avoid the possibility of certain weld defects

k) aluminium, though not included in the guide composition, is conventionally added in small amounts for restricting grain size growth in steels

l) residual elements are essentially the impurities in a refined steel. The steelmaker would be required to restrict them to as low a level as practicable though maximum permissible contents have also been specified in the guide composition as a general indication of the standard that could be reasonably expected.

C.2 Defect Acceptance Limits

C.2.1 Parent metal internal defect limits

The term "parent metal" is used for welded fabrications to refer to original component material as opposed to any weld metal connecting them. A guide defect specification for parent metal internal defect acceptance limits appropriate to forgings based on current standards for the nuclear power industry are as follows:-

Length : No defect to have a dimension greater than 20mm

Area : No defect to have a projected area in any direction greater than $40mm^2$.

Through-wall extent : Defects to account for no more than 3% of any wall thickness.

Circumferential extent : The total projected length of defects not to account for more than 2% of the outside circumference of a penetrator at any cross-section

Cracks : No defects identifiable as cracks or tears to be present

Note : If the clear distance between two otherwise acceptable defects is less than the square root of the sum of their areas then they are considered as a single defect equal in size to the least rectangle that can be drawn to enclose them both.

For a specification for steel castings to be viable it would have to be less restrictive than for forged components. It is, therefore, possible that slightly thicker wall sections would be required for castings to achieve similar structural and corrosion performances to forgings. However, this is by no means a definite consequence of the use of castings as penetrator behaviour models are insufficiently developed to allow a proper evaluation to be made at this time.

An appropriate guide defect specification for high integrity cast penetrator components based on current standards for offshore platforms has been identified as follows:-

Length	:	No defect to have a dimension greater than 50mm
Area	:	No defect to have a projected area in any direction greater than 250mm²
Through-wall extent	:	Defects to account for no more than 15% of any wall thickness
Circum ferential extent	:	The total projected length of defects not to account for more than 10% of the outside circumference of a penetrator at any cross section

The prohibition of cracks and note on defect separation from the specification for forged products would also be applied to castings.

C.2.2 Surface defect limits

Surface defects are essentially a special case of internal defects but size limits for them are specified separately because of the different methods by which they are detected. Appropriate requirements for guidance at this stage for assessing penetrator manufacturing options are considered to be as follows:-

a) crack or tear-like surface breaking defects not permitted

b) any adherences such as sand or scale not permitted

c) defects not to exceed 6mm in length nor 2mm in width or depth after any surface finish machining for forged products

d) defects not to exceed 15mm in length nor 5mm in width or depth after any surface finish machining for cast products

C.2.3 Weld metal internal defect limits

A guide defect specification for weld metal has been based on the recommendations of British Standard 5500 for high integrity pressure vessels as follows:

a) Planar defects : cracks, lamellar tears, lack of fusion and lack of root penetration not permitted

b) Cavities

i) isolated pores (or individual pores in a group) not to exceed 4.5mm in diameter

ii) uniformly distributed or localised porosity not to exceed 3% by area (as would be seen in a radiograph)

iii) linear porosity, aligned wormholes and surface cavities not permitted

iv) isolated wormholes and crater pipes not to exceed 6mm in length nor 1.5mm in width

c) Solid inclusions

i) Individual slag inclusions parallel to weld axis not to exceed 75mm in length nor 4mm width or height. (Inclusions to be separated by not less than the length of the longer inclusion). Sum of the lengths of inclusions not to exceed the total weld length

ii) Randomly oriented individual slag and tungsten inclusions (not parallel to weld axis) - as b i).

iii) Non linear groups of slag and tungsten inclusions - as b ii).

iv) Copper inclusions not permitted

d) Profile defects

i) Undercut permitted if intermittent and slight (depth not exceeding 0.5mm)

ii) Shrinkage grooves and root concavity permitted if intermittent and slight (depth not exceeding 1.5mm)

iii) Excess penetration not to exceed 3mm in height

iv) Reinforcement to blend smoothly with parent metal and shape not to interfere with non-destructive testing

v) Overlap not permitted

vi) Linear misalignment between components not to exceed 4.5 mm

APPENDIX D

DESIGN VARIATIONS

APPENDIX D

DESIGN VARIATIONS

D.1 Introduction

Some modifications to details of the reference penetrator design which have been considered during the course of the study are discussed in the following sections. Principal alternatives such as different wall thicknesses, materials and penetrator lengths have already been reviewed in 2.0.

D.2 Nose and Tail Profile Simplification

The nose and tail shapes of the reference design require a relatively large amount of material and are difficult to fabricate by any means except casting. All the alternatives to casting would involve extensive machining to remove surplus metal from an initial rough shaped steel ingot and this would be time consuming, wasteful and costly.

If further development of penetrator performance models showed that the reference design nose and tail shapes were not essential then there would be considerable manufacturing advantages by adopting the more rounded lighter options shown in Figure D1. The fabrication of such shapes is described in 6.4 and 6.5.

D.3 Welded Fins

Figure D1 shows fins welded to the penetrator tube rather than forming part of the tail as with the reference design. This would simplify the shape and welding of fins whilst also allowing shorter tail pieces to be used as they would not have to be of sufficient length to accommodate the attachment of fins. This would be a particularly valuable modification as long as hydrodynamic performance remained satisfactory and welding to the penetrator tube did not severely affect its structural or corrosion performance.

D.4 Mounted Fins

A scheme for mounting a set of fins on a penetrator tube without welding is also shown in Figure D1. This would avoid the potential problems of welding although the possibility of excessive local crevice corrosion would have to be checked. Reasons for discounting mechanical jointing such as screw coupling for nose to tube and tail to tube connections at this stage were given in 2.7.2.

D.5 Bimetallic Construction

The possibility of bimetallic construction was mentioned in 2.8.3 in terms of titanium clad steel penetrators. Similar schemes are being evaluated as candidate overpacks in certain geological disposal studies (Ref. 28).

The titanium cladding forms a corrosion resistant engineered barrier over the thick steel section resisting the applied loads. There would be many difficulties to be overcome such as the complication of welding bimetallic components.

There are some examples in current engineering practice of bimetallic pipework to combine corrosion resistance with strength. However, it is more frequently decided that components made entirely of the expensive corrosion resistant material will be cheaper overall when fabrication costs are included. Therefore, a more likely alternative to tube bimetallic construction is surface coating and this is discussed next.

D.6 Galvanising and Other Surface Treatments

A zinc galvanising coating applied to a steel penetrator would act as a cathodic corrosion protection for the penetrator as evident from Table 2.7.

The disadvantage of most surface coatings is that they are ineffective if they are damaged. Penetrators would be subjected to some surface abrasion during seabed impact and embedment so ruling out insufficiently robust coatings.

Galvanising, on the other hand, would continue to act as cathodic corrosion protection for a steel penetrator body even if damaged to the extent of complete removal from localised areas. Galvanising could, therefore, be an appropriate measure. For example, it could lead to significant steel weight savings and so avoid a possible difficulty in devising suitable handling equipment for very heavy penetrators.

D.7 Sacrificial Nose and Tail Attachments

The provision of sacrificial metal attachments would be an alternative cathodic corrosion protection system to galvanising. Zinc, aluminium or magnesium alloys would be candidate materials for this purpose.

Figure D1 shows sacrificial elements fitted over the simplified nose and tail profiles mentioned in D.2. This arrangement facilitates the fabrication of penetrator components. It also locates the sacrificial elements close to the ends of the penetrator tubes so affording most protection to the possibly vulnerable circumferential welds.

D.8 Instrument Hold Fabrication

Instrument hold fabrication is mentioned here in order to record the possibility that fittings other than the basic nose, tube, tail and fin components might be incorporated in penetrator designs. This would involve local features including areas of machining and welding which might affect a detailed evaluation of penetrator performance. For example, a badly designed attachment which left a narrow gap or crevice between itself and the penetrator body could initiate local corrosion when embedded in the seabed.

No definite size requirement has been specified for instrument holds as yet so nominal dimensions of 200mm diameter by 200mm length have been assumed. A welded formed plate instrument hold is shown attached, also by welding, to a penetrator tail in Figure D1. An alternative entailing machined recesses in penetrator tails is also shown in the Figure.

It is currently assumed that all penetrators would incorporate instrument holds but they would be relatively small and so have not been shown in any illustrations apart from Figure D1.

It should also be noted that actual penetrators might incorporate other features such as handling lugs even though none has been shown in any Figures at this stage.

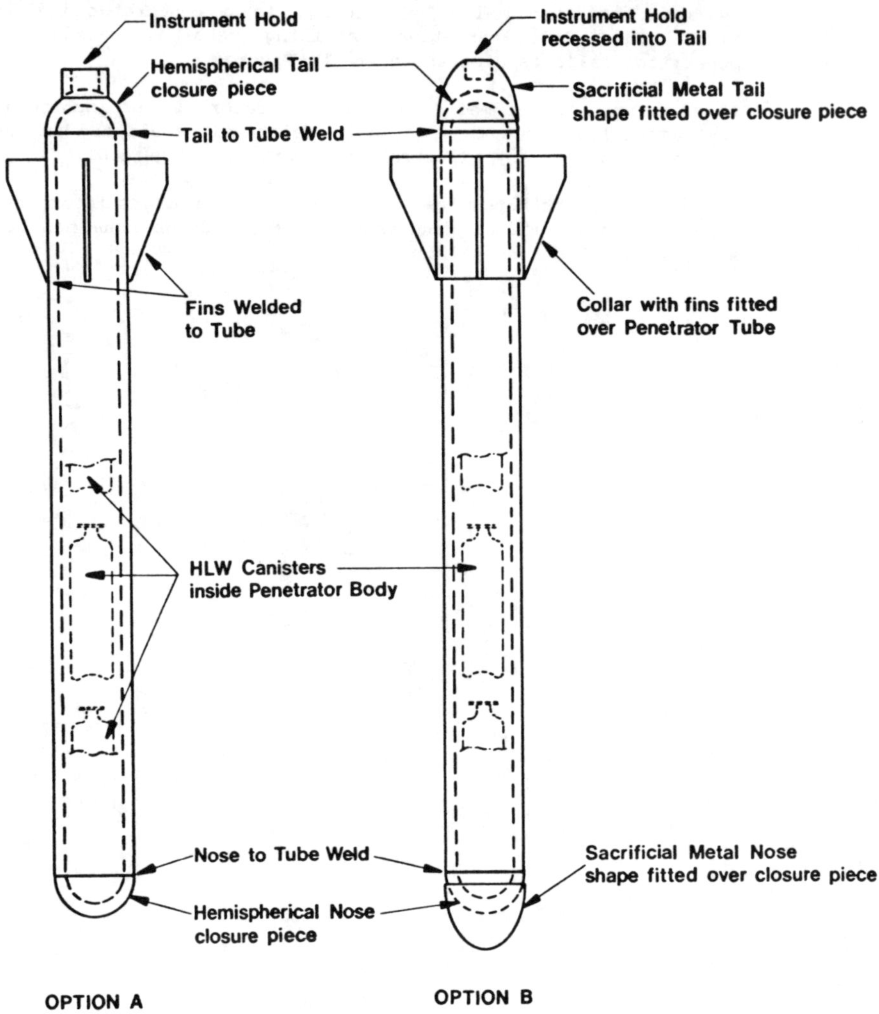

PENETRATOR DESIGN DEVELOPMENTS

figure D1

APPENDIX E

TUBE MANUFACTURE

APPENDIX E

TUBE MANUFACTURE

E.1 Tube Forging and Machining

Penetrator construction from machined forgings would be the most versatile but also one of the most expensive fabrication options.

Only solid forging has been considered in this study although hollow forging techniques do exist. In hollow forging the ingot is pierced and then forged around a die to form the bore. However, hollow forged tubes invariably require machining to give serviceable finishes which also allow adequately sensitive non destructive testing to be performed. Thus, as a rule, less than 600mm diameter bores are considered more economical to solid forge and bore out than to hollow forge and machine finish. Penetrator tubes fall in the former category because of their 500mm diameter bores. Furthermore, it is unusual to hollow forge wall thicknesses to less than about 150mm on large workpieces.

E.1.1 Process description

A typical production sequence is outlined in Figure 16 and described below:-

a) Ingot pouring: special ingots for high integrity products can be poured in special enclosures under a vacuum to avoid gas voids and oxidation.

b) Ingot end discarding: most metal impurities and faults concentrate at the bottom or float to the top of an ingot and are thus eliminated by discarding the ingot ends.

c) Step 1 forging: the temperature of an ingot is first raised into the hot working regime (which typically begins at about 60% of the absolute melting point) and forged to break down the coarse grained as-cast metallurgical structure to a workable, chemically homogeneous, wrought structure.

d) Rough forging: final or step 2 forging of many products is undertaken in the cold working regime (which typically ends at about 25% of the absolute melting points of steels) because of the reduced problem of non uniform cooling distortions. However, the relatively large penetrator body billets would have to be rough forged in the hot working regime using open dies as this would need less force and give the necessary degree of workability to achieve the required billet elongation. Rough forging would thus be a continuation of the step 1 forging. A 3000 - 4000t capacity press would be expected to be adequate for forging.

e) Reheating : it is likely that billets would have to be returned to a furnace at intervals during forging to maintain their temperature in the hot working regime and so reduce the possiblity of shear band and internal hole defects occurring.

f) Rough machining: the rough forged bar shape would be left with a machining allowance of about 40mm so that inevitable forging surface irregularities could be removed. This is done in a rough machining stage so that the forging can be checked for any defects before subsequent expensive heat treatments and final machining.

g) Heat treatment : the main heat treatment, normalizing in the case of penetrators, can give rise to distortions due to relaxation of internal stresses and so would be undertaken before final machining. Such distortions are reduced by vertical hanging of long slender components in the furnace. Hydraulic straightening rams are commonly used to remove any gross deformations after heat treatment and should not be prohibited unless a detrimental effect on say, penetrator corrosion mechanisms, was identified.

h) Finish machining : finish machining techniques are described in E.6.

i) Stress relief heat treatment : this final heat treatment could be delayed until after tube to tail fabrication welding.

j) Final NDE : non destructive examination techniques are described in 6.7.

k) Production rate : no meaningful estimate of press occupation time per tube is possible because of the "stop-start" nature of forging between furnace reheats of the billets to maintain their temperatures in the hot working regime (multi-heat forging). However, for guidance only on the timescales involved, one manufacturer felt that a weekly production rate of about five reference design penetrator tubes might be expected.

l) Note that tubes for the shorter penetrator options would be cut from long forged and machined lengths of upto about 20m as the most economic manufacturing procedure.

E.1.2 Process advantages

Relative advantages of machined forgings for penetrator tubes are:-

a) virtually no limits on tube length or wall thickness (ingots weighing in excess of 550t have been poured) and high dimensional accuracy possible with machining.

b) recesses, lugs, tapers or any other necessary handling features on penetrators can be readily formed during machining.

c) tube with integral nose fabrication possible.

d) good chemical homogeneity and metallurgical structure characteristic of forged products.

e) high purity defect free components readily produced in wide range of steel grades.

f) forgings have non-isotropic mechanical properties but the orientation of optimum strength, ductility and other parameters can be chosen to be of best advantage to penetrator performance.

E.1.3 Possible drawbacks

Possible drawbacks of machined forgings for penetrator tubes are:-

a) least "production-line" type of process so more expensive than centrifugal casting and other techniques developed especially for tube manufacturing.

b) relatively little control over production parameters. That is to say that factors such as forging times between reheats and press operation depend on the judgement of the equipment operators. Thus there is a certain "lack of reproducibility" between tubes in a production batch.

E.1.4 Production size range

Virtually the only limits on penetrator machined forgings are those determined by the size capacity of machine tools which are discussed in E.6.

In general, any tube upto about 20m long could be produced in a major machine shop. The only real difficulties are introduced when relatively thin wall thicknesses are required. Thus the maximum length for machining of 25mm wall thickness 500mm internal diameter tubes would be about 8m. These limits are set out in Table 6.1 and shown in Figure 22.

E.1.5 Penetrator specification satisfaction

Of all the tube manufacturing processes, machined forgings would have least difficulty in satisfying the guide penetrator specifications developed in 5.2. In particular, steel defect sizes would be small because of forging. Also machining would give very close tolerances and fine surface textures. A more detailed breakdown of specification satisfaction is given in E.5.5 for tube extrusion which is also a forging process.

E.2 Tube Solid Casting and Machining

Machined solid castings could be an appropriate compromise between machined forgings and hollow castings for penetrator construction. The solidification defect zone characteristic of casting can be caused to form along the central axis and thus be removed during the machining stage. The need for machining gives the same design versatility and tolerance control as for machined forgings and savings would result by casting the steel to the outside dimensions of penetrators without the need for costly forging and extensive external machining. However, the advantages of forging for metallurgical homogenisation and defect size breakdown would be lost.

Longitudinal axis vertical or horizontal casting would be equally appropriate and the choice between them would depend largely on the preference of the appointed foundry based on experience of similar projects. On balance though, horizontal casting would probably be preferred and especially so with regard to the longer penetrator design options. Therefore, only horizontal casting has been described.

E.2.1 Process description

A schematic hollow casting configuration is shown in Figure 17. An arrangement for solid casting would not have a sand core, intermediate feeders or padding. The casting would be inclined slightly upwards from the in-gate end to a riser at the far end. Thus slag, sand and other inclusions would tend to float to the riser rather than collect along the crown of the casting.

A typical solid casting sequence outline would be:-

a) Pattern making : hard wood patterns possibly with metal facings would be a common choice for the likely size and numbers of penetrators. Patterns are used to form the sand moulds described below and would be expected to have a considerable useful lifetime in excess of 100 mould forming cycles.

b) Mould making : modern foundry sands are predominantly chemical bonded. They are poured around the patterns mentioned above and left to cure. The curing of some sands relies on the use of carbon dioxide as a catalyst. The cured moulds are stripped from the patterns and their surfaces coated with a refractory wash to improve the finishes on and act as bond breakers for the castings. A sand mould would only be expected to be used for a few, possibly only one, castings given the likely size of penetrators. Used sand is recycled by re-mixing with chemical bonding agents for subsequent mould making.

c) Casting pit preparation : mould segments are arranged in a casting pit to build up a complete mould shape as shown in the Figure. Holding down beams are necessary to resist the buoyancy of the upper mould segments in molten steel.

d) Steel pouring : mould filling would take around 15 minutes at typical steel input rates for the likely size of penetrator castings.

e) Mould stripping : parting lines between mould segments are arranged to facilitate stripping for removal of the castings.

f) Heat treatment : cooled castings have coarse brittle grain structures and are thus often heat treated, normalising in the case of penetrators, before feeder head removal and other operations which carry the risk of cracking the castings. The remarks in E.1.1.g on mechanical straightening of forgings also apply to solid castings.

g) Fettling : feeder heads and the like are removed and shot blasting or other processes used to remove adhering sand and surface blemishes. An initial inspection for casting defects is then possible before subsequent machining.

h) Finish machining : finish machining techniques are described in E.6.

i) Stress relief heat treatment : this final heat treatment could be delayed until after tube to tail fabrication welding.

j) Final NDE : non destructive examination techniques are described in 6.7.

k) Production rate : in very general terms and for guidance only to the timescales involved, one manufacturer felt that a weekly production rate of about ten reference design penetrator tubes might be expected. A period of about 20 hours per casting for mould assembly, steel pouring and mould stripping in the foundry casting pit was envisaged.

E.2.2 Process advantages

Relative advantages of machined solid castings for penetrator tubes are:-

a) virtually no limits on tube length or wall thickness (castings weighing in excess of 300t are quite feasible) and high dimensional accuracy possible with machining.

b) recesses, lugs, tapers or any other necessary handling features on penetrators can be readily formed during machining.

c) tube with integral nose fabrication possible.

d) sand castings can be made with sufficiently good surface finishes to meet non destructive testing requirements after shotblasting during fettling without the need for the extensive external machining necessary on rough forgings.

E.2.3 Possible drawbacks

Possible drawbacks of machined solid castings for penetrator tubes are:-

a) not a particularly "production-line" type of process (though more so than rough forging) and thus at a possible disadvantage to centrifugal casting and other techniques developed especially for tube manufacturing.

b) difficult to have full control over production parameters. That is to say that factors such as pouring rates and molten steel flow behaviour can vary for different castings. Thus there is a certain "lack of reproducibility" between tubes in a production batch.

c) castings are prone to chemical segregation during solidification. For example, steel in one area can have a marginally richer carbon content than steel in another area of a casting.

d) various types of grain growth occur at different locations, generally depending on their position relative to a mould wall, during solidification. Thus a casting can exhibit a chill zone immediately adjacent to a mould wall followed by columnar dendritic, branched dendritic, equiaxial dendritic and non-dendritic grain structures at increasing distances from the mould wall.

e) heat treatments such as normalizing or annealing have a limited effect on balancing chemical segregation or transforming the initial solidification orientation of grain structures of castings. Instead, heat treatments are necessary to break down or refine the coarse brittle as-cast structure by stimulating renewed grain formation to give finer more ductile structures. However, castings are no different to forgings in requiring heat treatments to achieve suitable combinations of mechanical properties.

f) considerable skill is necessary for mould design and casting procedure specification to avoid the development of hot tears, micro-porosity and other solidification shrinkage defects in critical areas of castings.

g) mould sands and their bonding agents give off gases when in contact with high temperature molten steel. Moulds are designed to avoid so called hot spots and to allow any gases to vent without forming void defects known as blowholes in castings. Nevertheless, a certain degree of gas uptake by the steel is inevitable and is generally held to make vacuum degassing superfluous as regards casting.

h) moulds and casting procedures are designed to avoid slag and loose sand inclusions but some defects of this nature are, nevertheless, a feature of casting. Such defects in ingots are broken up by forging for wrought steel products but this solution is not, of course, applicable to castings: Consequently, less severe internal defect acceptance levels are specified for castings than for forgings as discussed in 5.2.7. Design codes also recognise this factor by adopting lower working stresses for castings.

E.2.4 Production size range

Similar size limitations apply to machined solid castings as to machined forgings in that machine tool capacity is virtually the only restriction.

In general, any tube up to about 20m long could be produced in a major machine shop. The only real difficulties are introduced when relatively thin wall thicknesses are required. Thus the maximum length for machining of 25mm wall thickness 500mm internal diameter tubes would be about 8m. These limits are set out in Table 6.1 and shown in Figure 22.

E.2.5 Penetrator specification satisfaction

Machined solid castings would generally satisfy the penetrator specifications developed in 5.2 though this statement can be expanded upon as follows:-

(a) Chemical composition : a certain degree of chemical inhomogeneity can occur during solidification in the foundry mould.

b) Mechanical properties : the guide mechanical properties proposed for penetrators can be readily achieved. However, pressure vessel design codes recommend that working stresses in statically cast components should be restricted to between 70 and 90% of those permitted in equivalent forgings because of the possibility of larger defects in casting.

c) Grain size : grain refinement is achieved by appropriate heat treatments although the various dendritic grain orientations which form at different distances from mould walls during solidification cannot be greatly altered.

d) Tolerances : machining would give very close tolerances although fettled castings before final machining would also be expected to satisfy the guide tolerances proposed in 5.2.5 for penetrators.

e) Surface texture : machining would give very fine finishes although unmachined fettled castings might possibly be sufficient to meet non destructive testing requirements.

f) Steel defects : practical limits on defect sizes are greater, that is less satisfactory, for castings than for forgings as discussed in 5.2.7.

E.3 Tube Hollow Casting

Tubes are, in fact, one of the most difficult shapes to cast. A plain sided tube, that is a tube without wall thickness tapers, is virtually impossible to cast in any size without the formation of internal defects during solidification. There are, however, techniques such as tapering or padding of walls for controlling the progression of solidification to avoid defects and these are described in this section. An obvious drawback is that machining becomes necessary to remove unwanted tapers and padding from tubes for construction of penetrators.

Longitudinal axis vertical and horizontal casting techniques exist for tubes. Vertical casting, however, entails wall tapering in the region of 1:20 or a significant feeding head in the region of 30% of the tube height or both to avoid major solidification defects. On balance, therefore, horizontal casting would probably prove more suitable particularly for the longer penetrator design options. Consequently only horizontal casting has been described.

E.3.1 Process description

A schematic hollow casting configuration is shown in Figure 17. A typical solid casting process was described in E.2.1. Hollow casting would follow a similar sequence so only its special aspects are outlined below:-

a) Core making : the tube bore would be formed by a sand core. Some, but by no means all foundries prefer cores with tapers in the region of 1 : 200 to facilitate stripping from a core pattern box. Otherwise, parallel sided cores can be formed in split pattern boxes or by strickling to shape. Sand cores are readily broken out from solidified castings. The core has to be reinforced to span the length of the tube under its own weight when the mould is empty and under even greater buoyancy effects when the mould is filled with molten steel. The possible use of chaplets to support very long cores is discussed in (d).

b) Padding : local thickening or padding of the tube wall is necessary to encourage solidification to begin at the bottom and progress to the top of the mould. In this way, molten steel in the top of the mould can flow to take up shrinkage as solidification progresses upwards.

c) Riser/feeders : reservoirs or riser/feeders are necessary at regular intervals along the tube crown as sources of molten steel to make good shrinkage as the casting solidifies.

d) Chaplets : the maximum feasible span for a 500mm diameter core is around 8m. Intermediate supports known as chaplets, typically 20mm diameter metal props between the outer foundry mould and the core, are necessary for longer hollow casting lengths. Chaplets become cast into the tube walls and subsequently have to be drilled out and replaced with weld deposited metal or, if adequately fused, be left in place with only local surface blending repairs made.

Tube with integral nose casting is also possible by using chaplets to support the free end of an internal core. In general, however, it is not considered that chaplets would be appropriate because of the possibility that they might create preferential corrosion paths through penetrator bodies.

E.3.2 Process advantages

Relative advantages of hollow casting for penetrator tubes are:-

a) virtually no limit on tube wall thicknesses that can be cast. Also tube lengths in excess of 8m are possible if the use of chaplets is allowed although this is unlikely for the reason given in E.3.1.d.

b) a significant amount of external machining is necessary to remove unwanted padding but this has the advantage of giving high external dimensional accuracy. Also recesses, lugs, tapers or any other necessary handling features on penetrators can be readily formed during machining.

c) tube with integral nose fabrication is possible if the use of chaplets is allowed. However, this is unlikely for the reason given in E.3.1.d.

E.3.3 Possible drawbacks

All the possible drawbacks of solid casting listed E.2.3 also apply to hollow casting.

The fundamental difference between solid and hollow penetrator tube casting is that molten steel in the core of a solid casting feeds the solidification shrinkage of the outside of the casting whereas external padding and feeder head arrangements are used for hollow casting. Centreline shrinkage defects occur in solid castings but are of no consequence because the core of the casting is machined out to create a tube. External machining is necessary to remove unwanted padding which prevented the formation of shrinkage defects in the wall thickness of hollow castings.

It should be noted that some forms of internal machining could still be necessary for hollow casting to give a satisfactory surface texture for non destructive testing requirements.

E.3.4 Production size range

Maximum feasible sand core span limits the possible length of hollow castings to about 8m assuming chaplets are not used for the reason given in E.3.1.d. In general, casting wall thickness limits are only governed by machine tool restrictions. Thus 25mm would be about the thinnest wall that could be machined on an 8m long tube. These limits are set out in Table 6.1 and shown in Figure 22.

E.3.5 Penetrator specification satisfaction

The specification satisfaction considerations of solid castings listed in E.2.5 also apply to hollow castings. However, it should be noted that hollow castings would generally require more extensive external machining than solid castings. Therefore, the remarks regarding fettled solid castings with respect to tolerances and surface texture would not be relevant to hollow castings.

E.4 Tube Centrifugal Casting

Centrifugal casting is uniquely applicable to tube fabrication for which it is particularly effective. It is the closest approach to a production line process especially for relatively small quantities of tubes with unusual wall thickness to diameter ratios.

Tubes are formed directly and so require very little or even no surface machining depending on their intended applications. Centrifugal casting eliminates most of the metallurgical defects characteristic with static sand casting and should, as such, be considered as a process in its own right rather than a variation of conventional static casting.

It is felt that centrifugal casting, together with pipe extrusion discussed in E.5, would be eminently suitable for fabrication of the thicker walled penetrator tube options.

E.4.1 Process description

A schematic centrifugal casting bed is shown in Figure 18. A typical production sequence would be as follows:-

a) Mould spinning : the mould rests on and is rotated by driven rollers in the casting bed. Larger casting facilities sometimes include damping rollers to reduce harmful vibration effects.

b) Mould preparation : the mould is fitted with refractory end plates, pre-heated and sprayed inside with a mould wash which dries to give a ceramic coating of about 1mm thickness. The ceramic coating prevents premature solidification when molten steel is subsequently introduced to the mould and then forms a bond-breaker between the centrifugally cast tube and the mould.

c) Sand lining : an alternative mould preparation technique sometimes employed for larger castings entails the insertion of preformed foundry sand liners into the mould. The inside of the sand liner rather than the mould itself being coated with a refractory wash. Varying external diameter tubes such as required for underwater pipeline buckle arrestors can also be made by appropriate profiling of sand liners. The use of sand liners has not been considered for penetrator fabrication as they can cause a poor external surface zone on tubes.

d) Centrifugal casting : the mould is rotated whilst a pre-determined weight measure of molten steel is introduced through a central hole in an end plate. The molten steel is spun into a tube shape before it solidifies. 95 to 120 times gravity centrifugal forces float non-metallic inclusions, which are less dense than the molten steel, to the inside surface of the tube. It is also likely that the high centrifugal forces are effective at eliminating gas microporosity in the tube body. Therefore , manufacturers have suggested that vacuum degassing is unnecessary for centrifugal castings.

e) Special procedures : Figure 18 shows a casting bed with a cooling duct and sliding enclosure. These are special and relatively novel features developed for particularly high integrity production. The sliding enclosure when used with an appropriate pouring chute allows casting to be undertaken in an inert gas purged atmosphere to prevent oxidation of the molten steel. The cooling duct can be used to accelerate the solidification of the steel once it has been spun into a tube. It induces steep radial temperature reduction gradients across the tube wall to the mould body. These generally give beneficial metallurgical grain structures and confine shrinkage defects such as porosity to only about 10% of the wall thickness adjacent to the inside surface of the tube.

f) Tube extraction : cooling contraction of the solidified tube allows it to be withdrawn from the mould.

g) Shotblasting : tubes are shotblasted to remove any adhering ceramic coating and enable an initial inspection before further processing is undertaken.

h) Heat treatment : the main heat treatment, normalizing in the case of penetrators, can give rise to distortions due to relaxation of internal stresses and so would be undertaken before final machining. An initial slight bowing sometimes occurs during cooling of tube immediately after it is extracted from the casting mould. Such distortions are reduced by vertical hanging of long slender components in the furnace. Hydraulic straightening rams are commonly used to remove any gross deformations after heat treatment and should not be prohibited unless a detrimental effect on, say, penetrator corrosion mechanisms, was identified.

i) Finish machining : tubes are pull-bored out to remove internal shrinkage defect zones and outside machined if necessary to satisfy a fine surface texture specification.

j) Stress relief heat treatment : this final heat treatment could be delayed until after tube to tail fabrication welding.

k) Final NDE : non destructive examination techniques are described in 6.7.

l) Production rate : a typical casting bed production cycle requires about 30 minutes per tube though many other factors affect overall production times.

E.4.2 Process advantages

Relative advantages of centrifugal castings for penetrator tubes are:-

a) Large wall thickness to diameter ratio tubes readily produced in a wide range of steel grades to good dimensional accuracy particularly with respect to concentricity.

b) Profiled foundry sand liners could be used in moulds to give varying external diameter tubes if required to accommodate, for example, penetrator handling lugs.

c) uniform and isotropic mechanical properties throughout wall thickness.

d) high metallurgical cleanliness in finished tubes because defects float during spinning to inside face which can then be pull-bored out.

e) external surface texture can be sufficiently fine to meet non destructive examination requirements after shotblasting without the need for machining.

f) economically produced in small quantities to meet individual material and dimension specifications.

g) comprehensive control over production parameters possible so that high degree of similarity or reproducibility between tubes in a production run can be achieved.

h) good control over grain size growth can be achieved by the use of mould cooling or casting in high pressure atmospheres or both. Further grain refinement is possible with subsequent heat treatments.

E.4.3 Possible drawbacks

Possible drawbacks of centrifugal castings for penetrator tubes are:-

a) possibility of chemical segregation during spinning. For example, steel at one wall depth can have a marginally richer carbon content than steel at another wall depth.

b) various types of grain growth occur at different distances from the mould wall during solidification. A much thicker external surface zone of columnar dendritic grain structure is characteristic of centrifugal castings as opposed to static sand castings although this is not generally detrimental.

c) hot tears can occur on outside tube surfaces during solidification if unsuitable refractory mould washes or production parameters are chosen.

d) defects can be caused by casting bed vibrations. Necessary vibration damping precautions become increasingly difficult and sophisticated for large castings and limit the lengths and, more fundamentally, the weights of tubes that can be made.

E.4.4 Production size range

Equipment vibration problems limit the weight of tubes that can be centrifugally cast. Facilities in Japan are reported to have upto 50t tube making capacities but the largest casting beds known to be in use in Europe are only rated for around 12t production. This puts the European centrifugal casting capacity on a par with the production size range for tube extrusions discussed in E.5. Also, 12t capacity equipment has generally been developed for very high quality tube production for special pipeline and process industry application.

50t capacity equipment has been used to fabricate large paper rolls and structural columns for bridges and buildings when size has been more important than exceptional metallurgical quality. Therefore, at this stage, it is considered reasonable and conservative to assume that a finished tube weight of 12t represents the maximum capacity of equipment that would be suitable for penetrator tube fabrication.

General production size range limits can then be given as:-

a) 12t maximum weight
b) 1.4m maximum external diameter
c) 9m maximum length
d) 10mm minimum wall thickness (or greater if needed for tube handling rigidity)
e) maximum wall thickness after removal of internal shrinkage defect zone = one third of outside diameter.

These limits are set out in Table 6.1. and shown in Figure 22 from which it is possible to read the maximum length of 500mm internal diameter tube that can be centrifugally cast with a particular wall thickness. Thus the maximum length for 75mm wall thickness tube is about 9m.

E.4.5 Penetrator specification satisfaction

Centrifugally cast tubes would readily satisfy the penetrator specifications developed in 5.2 though this statement can be expanded upon as follows:-

a) Chemical composition : a certain degree of chemical segregation or inhomogeneity can occur during tube spinning in centrifugal casting moulds.

b) Mechanical properties : the guide mechanical properties proposed for penetrators in 5.2.2 can be readily achieved. Properties such as elongation tend to be better than for static sand castings but marginally less than for forgings in the same material. However, in contrast to forgings, centrifugal castings have uniform isotropic properties which are less affected by welding.

c) Grain size : grain refinement is achieved by production parameter control and appropriate heat treatments although the dendritic grain orientations which form during solidification remain essentially unaltered by the heat treatments.

d) Tolerances : machining would give very close tolerances although shotblasted centrifugal castings would satisfy the external diameter guide tolerances proposed in 5.2.5 for penetrators. There is less control over as cast internal diameters because of molten steel weighing inaccuracies. However, the internal shrinkage defect zone would almost certainly have to be machined out for penetrator tubes so giving close finished internal diameter tolerances.

e) Surface texture : machining would give very fine finishes although mould ceramic wash behaviour gives centrifugal castings characteristic finely-dimpled outside surfaces with approximately BS 1134 Ra =25μm being only one grade rougher than the 12.5μm guide value proposed in 5.2.6. Thus it is possible that only shotblasting of external surfaces would be necessary for penetrator tubes if the proposed guide texture proved to be a conservative requirement.

f) Steel defects : practical limits on defect sizes are similar to those specified for forgings as discussed in 5.2.7.

E.5 Tube Extrusion

Extrusion is as elegant an approach as centrifugal casting to penetrator tube production. The fundamental difference is that extrusion gives high integrity forged, or wrought, steel structures as opposed to the cast structures of centrifugally spun tubes.

Extrusion closely approaches a production line process for relatively small quantities of tubes with large wall thickness to diameter ratios. Tubes are formed directly and so require minimal surface finishing.

It is felt that extrusion, together with centrifugal casting discussed in E.4, would be eminently suitable for fabrication of the thicker walled penetrator tube options.

E.5.1 Process description

A typical production sequence is outlined in Figure 19 and described below:-

a) Ingot production : the pouring and preparation procedure described in E.1.1.a and b for open die forging also applies to ingot production for extrusion.

b) Ingot forging : the ingot temperature is first raised to around 1150°C, forged or upset once to dislodge scale and placed in a cylindrical blocking pot. The ingot is then forged in the blocking pot to become a squat cylindrical blocker. This ingot forging stage breaks down the course grained as-cast metallurgical structure to a workable, chemically homogeneous, wrought structure.

c) Blocker piercing : a hole of the same diameter as required for the bore of the extruded tube is formed in the centre of the blocker with a piercing punch.

d) Tube extrusion : whilst still red hot, the hollow blocker is placed on a pedestal in a high capacity, typically 270 to 320 MN (30 to 35,000t) hydraulic press. An internal punch is inserted through the blocker to maintain the internal bore and then the press is operated to lower an extrusion throat onto the blocker and force it to extrude vertically upwards into a tube. Blocker to tube wall thickness reduction ratios typically range between 5 and 10 : 1.

e) Surface finishing : internal and external surfaces are ground to give the required finish and allow a check for any defects before subsequent processing.

f) Heat treatment : the main heat treatment, normalizing in the case of penetrators, can give rise to distortions due to relaxation of internal stresses and would be undertaken before final machining. Such distortions are reduced by vertical hanging of long slender components in the furnace. Hydraulic straightening rams are commonly used to remove any gross deformations after heat treatment and should not be prohibited unless a detrimental effect on, say, penetrator corrosion mechanisms, was identified.

g) Stress relief heat treatment : this final heat treatment could be delayed until after tube to tail fabrication welding.

h) Final NDE : non destructive examination techniques are described in 6.7.

i) Production rate : a typical extrusion press production cycle requires about 15 minutes per tube though many other factors affect overall production times.

E.5.2 Process advantages

Relative advantages of extrusion for penetrator tubes are:-

a) large wall thickness to diameter ratio tubes readily produced to good dimensional accuracy.

b) profiled extrusion throats can be used to extrude non circular tubes. Thus, for example, longitudinal ribs could be left on the outside of tubes if required to accommodate penetrator handling lugs.

c) good chemical homogeneity and metallurgical structure characteristic of forged products.

d) high purity defect free tubes readily produced in wide range of steels.

e) forgings, such as extrusions, have non-isotropic mechanical properties but the longitudinal orientation of optimum strength, ductility and other parameters could be advantageous for penetrator performance.

f) external and internal surface textures need only minimal finishing by grinding to meet non destructive testing requirements.

g) economically produced in small quantities to meet individual material and dimension specifications.

h) reasonable control over production parameters possible so good similarity or reproducibility between tubes in a production run can be achieved.

E.5.3 Possible drawbacks

No distinct technical drawbacks have been identified to the use of extrusions for penetrator tubes. One aspect regarding pipe concentricity is discussed below, but it is not suggested that it would necessarily be significant. The apparent lack of disadvantages for extrusion does not, however, make it the automatic choice for penetrator construction as this depends more on comparing the relative advantages of candidate processes.

Wall thickness tolerances quoted for pipe extrusions are mainly to allow for unavoidable lack of concentricity of the bores. The lack of concentricity is greatest, but still within tolerance limits, at the first end that is extruded but invariably settles down to give a more concentric cross-section after the first 2m or so length of pipe. Production experience experience has shown that most pipe is extruded to well within the tolerances stated by manufacturers. The slight lack of concentricity is not a problem for the great majority of applications but if a fully concentric finished tube is required then an appropriate machining allowance can be included at the extrusion stage.

E.5.4 Production size range

Only one manufacturer of thick section tube extrusions has been identified. The following production size range information relates to the current facilities of that manufacturer.

There is a 30,000t capacity multi-purpose press which is used to produce pipe extrusions in Scotland. There is, however, a 35,000t capacity press in America specifically for pipe extrusions which can produce roughly 25% heavier and longer pipes.

Approximate production size range limits for the Scottish press are:-

a) 12t maximum finished tube weight
b) 1.45m maximum external diameter
c) 13m maximum finished length
d) 20mm minimum wall thickness
e) 150 to 970mm internal diameter range though current internal surface finishing equipment cannot be used for less than 215mm diameter bores.

These limits are set out in Table 6.1 and shown in Figure 22 from which it is possible to read the maximum length of 500mm internal diameter tube that can be extruded with a particular wall thickness. Thus the maximum length for 75mm wall thickness tube is about 7m.

E.5.5 Penetrator specification satisfaction

Extruded tubes would readily satisfy the penetrator specification developed in 5.2 though this statement can be expanded upon as follows:-

a) Chemical composition : good chemical homogeneity is achieved with extrusion because it is a forging process.

b) Mechanical properties : the guide mechanical properties proposed for penetrators in 5.2.2 can be readily achieved. Forgings have non-isotropic mechanical properties but the orientation of optimum strength and other parameters would probably be advantageous for penetrator performance.

c) Grain size : grain refinement is readily controllable by appropriate selection of the degree of working or forging and subsequent heat treatments.

d) Tolerances : normal manufacturing accuracies of -0, +3.2mm on internal diameter and -0, +6.4mm or -0, + 10% if greater on wall thickness for tube extrusions are well within the tolerances proposed in 5.2.5 for penetrators.

e) Surface textures : only minimal grinding is necessary on tube extrusions to give surfaces with smoother textures than the BS 1134 Ra = 12.5μm guide value proposed in 5.2.6.

f) Steel defects : practical limits on defect sizes are equal or better than the guide specification applicable to forgings proposed in 5.2.7.

E.6 Tube Fabrication By Miscellaneous Processes

At this stage there is considerable uncertainty with regard to performance requirement setting and behaviour modelling for penetrators. This is particularly apparent from the wide variation, 25 to 200mm, of steel tube wall thicknesses that has been considered in the study pending the results of further research and development work. The processes described in E.1 to E.5 can all be employed to fabricate seamless tubes with wall thicknesses over the full 25 to 200mm range. However, there are other processes, some of which are very commonly used in the engineering industry, for tube fabrication over a more restricted wall thickness range. These less versatile processes are outlined in this section in order to give a comprehensive review of penetrator construction options.

E.6.1 Plate forming and seam welding

Seam welded tube is a particularly important option because of its economy and is the automatic choice for a wide range of engineering applications. Strip mill rolled plate is used so giving tubes the advantageous characteristics of forgings. In particular, the guide specifications developed in 5.2 for penetrators could be satisfied. Furthermore, there is no reason at this stage to object specifically to welding of the longitudinal seams on the grounds of possible detrimental behaviour in the seabed sediments. This topic requires confirmation by detailed research. However, it should be remembered that readily weldable steels are currently under consideration to facilitate nose to tube circumferential welding during final penetrator assembly.

Maximum wall thickness is the main limitation to the applicability of seam welded tubes. Plates thicker than about 40mm could not be formed without significant strain damage into the nominal 500mm internal diameter tubes proposed for penetrators. Nevertheless, it is quite plausible that fully developed radiological assessments and corrosion models could identify an appropriate penetrator steel wall thickness of under 40mm. The production size range for seam welded formed plate tubes is give in Table 6.1 and shown in Figure 22.

Figure 20 shows a typical "U" - "O" press tube fabrication sequence. "C" presses are an alternative to "U" - "O" presses but do not reduce the number of steps in tube production. Three roll bending would be another candidate plate forming process for relatively short tubes.

E.6.2 Pipe milling

Several methods such as the Pilger process have been developed for large scale seamless pipe manufacture or milling. In effect, they are production lines for continuous pipe forging. Nominal 500mm internal diameter pipes with up to 100mm thick walls are included in standard mill pipe production ranges as indicated in Table 6.1 and Figure 22.

It is thought unlikely that milled pipe would prove suitable for penetrator tubes. It is only economical to mill very large consignments for use as pipelines. The relatively small quantities that would be needed for penetrators would have to be selected from standard stock sizes of pipe. There would be very little if any flexibility to individually specify materials, qualities or sizes to satisfy special penetrator performance requirements. Even if a particular pipe size was selected there would be the danger of it being withdrawn from the standard milling range during the design life of penetrator disposal operation facilities such as assembly plants and ships. Handling and assembly equipment might then have to be expensively modified to suit a new pipe size.

Continuous casting is another production line type of process which would appear to be an unsuitable penetrator tube fabrication option for similar reasons to those given for rejecting pipe milling at this stage.

E.6.3 Shape welding

Shape or build-up welding is a special process developed by one manufacturer for very large thick walled pressure vessel fabrication. It is claimed to be a low capital cost method which overcomes the technical drawbacks of massive conventional forging. Narrow beads of weld metal are continuously deposited around a base or start-up tube until the necessary wall thickness is achieved. However, the materials, dimensions and numbers currently envisaged for penetrator bodies would not justify this fabrication option.

E.6.4 Machine tool capacities

Very large and sophisticated machining equipment has been developed to meet requirements for components for power stations, process plants, shipbuilding and other heavy industries. Major suppliers have extensive facilities capable of undertaking a wide range of machining operations.

As an example, one typical large manufacturer has general machine tools for up to 200t weight, 40m long, 3m diameter workpieces and for drilling up to 2m diameter, 20m deep holes. Larger equipment is available for specific applications. This demonstrates that all the size variants proposed for penetrators are well within the limits of current machining capacity.

E.6.5 Minimum wall thickness machining

Machining to leave a 25mm wall thickness on 500mm internal diameter tubes is possible but requires special precautions and monitoring to guard against collapse of the wall under cutting tool forces. The problem would probably be manageable for up to 8m long tubes but could become critical and reflect in the cost of any longer tubes.

E.6.6 Penetrator specification satisfaction

Machining would readily satisfy relevant penetrator specifications developed in 5.2 as follows:-

a) Tolerances : machining tolerances can be many times better than the guide values proposed in 5.2.5 for penetrators. Fine machining is expected to give better than ± 0.3mm accuracy on nominal dimensions of upto 1000mm. Therefore, coarse or rough machining which would give accuracies of ±2mm on nominal dimensions upto 1000mm would be more than satisfactory for penetrators as it is felt that a ±8mm external diameter tolerance would be acceptable. Straightness tolerances for penetrators would be most readily achieved if machining were undertaken after the main heat treatment of Normalizing. It would then be possible to eliminate the distortions that usually occur during heat treatments by mechanical straightening or machining or a combination of both.

b) Surface texture : the BS 1134 Ra = 12.5 μm guide surface texture specification suggested in 5.2.6 for penetrators is, in fact, not a very demanding requirement and could be readily achieved by any machining. Rough machining would be expected to give at least a one grade better finish of Ra = 6.3 μm whilst fine machining could typically give a further three grade improvement to Ra = 0.8 μm.

E.6.7 Bore trepanning

The fastest and most appropriate approach to straight through boring for penetrator tubes would be trepanning as shown schematically in Figure 21.

E.6.8 Tube with integral nose machining

Trepanning is generally not practicable for blind hole machining which would be necessary for tube with integral nose fabrication from solid forgings or castings. This is because of the difficulty of breaking the neck of trepanned cores at the blind ends of holes.

Therefore, a blind hole upto 2000mm in diameter is achieved by making a 75 to 200mm diameter pilot bore and then drilling progressively larger holes until the required diameter is reached. The maximum pilot hole diameter and maximum size increment of subsequent bores is limited by the area of steel that can be cut without causing excessive vibrational instability of the cutting tool. However, a special type of tool called a spade which has stepped diameter increases enables several hole size increments to be drilled simultaneously.

Blind hole drilling is shown schematically in Figure 21 which also indicates the relative difficulties of the options for blind hole end profiling.

APPENDIX F

WELDING PROCESSES

APPENDIX F

WELDING PROCESSES

F.1 Introduction

Detailed accounts of basic principles and processes are not included in this report as there are many examples of standard reference works on welding. The notes in this section only cover aspects of special relevance to penetrator construction.

F.2 Arc Welding

Individual arc welding processes of SAW, MIG, TIG and ESW are considered in subsequent sections. However, the following remarks are generally applicable to arc welding.

a) Parameter control : the increasing adoption of solid state electronics is causing a revolution in the control of welding parameters such as arc voltage and current so that improved weld quality, reproducibility and defect monitoring is being achieved.

b) Welding speed : as a general rule welding speed is inversely proportional to weld integrity. Thus the increased productivity and lower costs of multi-power, or wire, welding heads must be balanced against the possible quality reduction over the use of a single wire head. Similar considerations apply to the selection of optimum weld filler wire diameter with smaller sizes being preferred for high integrity welding. Arc welding speed calculations in this study have been based on a relatively modest weld pass deposition rate of 5 kg/hr. This is equivalent to laying a 5mm diameter weld bead at 0.6m/minute.

F.2.1 Submerged arc welding (SAW)

(a) 200mm thick welds are feasible with SAW but greater defect avoidance measures such as anti-tramp rolls are needed than for more common place thinner welds.

(b) SAW often gives poor quality root passes. Thus MIG or preferably TIG is usually specified for root passes prior to completion by SAW of joints which could be subjected to fatigue loading. However, since there is little likelihood of significant cyclic loads on penetrators it is noted that SAW root passes against inert backings would probably be satisfactory as the presence of any stress raising defects would not be critical.

(c) Heat build-up problems would be associated with penetrator circumferential welding. For example, 400°C temperatures have been reached after 6 passes of a 1m diameter 25mm wall thickness pipe weld. Molten metal weld pool size generally becomes uncontrollable above 300-400°C parent metal interpass temperatures. Liquid or gas cooling would help but it is likely that regular interuptions in welding would still be needed to keep temperatures below acceptable limits. Problems would then be experienced by the need to grind out the defect left where welding was temporarily stopped. It is advisable to specify automatic mechanical grinding for this as other defects can be accidently introduced when using hand held grinders, or linishers. Automatic grinding would of course be a pre-requisite for assembly welding.

F.2.2 Metal inert gas (MIG) and hot wire Tungsten inert gas (TIG) welding

(a) The silica deposit problem associated with MIG may be overcome in the near future by novel consumable formulations but TIG would probably still be preferred for weld quality assurance especially since the development of hot wire TIG now gives acceptable weld deposit rates.

(b) Hot wire TIG entails electrical resistance pre-heating of the weld filler wire such that 12-14 kg/hr deposition rates have been achieved. It is currently being developed for high integrity applications such as in nuclear power plants.

(c) MIG and TIG welding parameters are more readily adjustable to control weld pool size so there would be reduced heat build up problems compared with SAW.

F.2.3 Narrow gap SAW, MIG and TIG

(a) Narrow gap welding processes are being developed for their economy. Reduced times for welding are necessary though greater equipment and workpiece setting-up accuracy is needed.

(b) Defects are difficult to rectify if they are not detected as they occur so that welding can be interrupted to allow their repair.

(c) Heat build-up is less of a problem than for conventional welding.

(d) Special measures are needed to ensure side wall fusion. The Japanese use mechanical wire twisting or flapping but others are developing electronically controlled pulsed arc systems which are giving promising results.

(e) Early welding heads and weld gap designs had to be modified so that the narrowing of a weld gap due to weld heat input distortions did not trap the welding head.

(f) Patent applications have been made by Burmeister and Wain Diesel A/S of Denmark for narrow gap welding of ship propellor shafts with roughly similar wall thicknesses and diameters to those envisaged for penetrators.

F.3 Electroslag Welding (ESW)

(a) It is not felt that ESW would give suitable weld qualities for application to penetrators.

(b) It has been suggested that ESW could be developed for use with cast irons but it is noted that very high temperature and prolonged heat treatments would be entailed.

F.4 Electron Beam Welding (EBW)

(a) A large EBW unit has been developed and installed by the Welding Institute. It is rated at 75 kW, has a 100m³ vacuum chamber and has made 300mm deep welds in steel. Thick welds have also been successfully accomplished in other materials such as 100mm of copper. Setting up a workpiece in the chamber typically takes 20 to 30 minutes followed by about 45 minutes evacuation time to establish the necessary vacuum. 85mm deep steel is welded at 150mm/min with 39 kW power input. 280 mm deep steel is welded at 50mm/min with 75 kW power input.

(b) The longitudinal axis of a penetrator would have to be vertical for circumferential EBW.

(c) A joint gap of not more than 1mm is usually specified but gaps of upto 2mm are tolerable.

(d) Magnetic lenses induce beam spinning to make the effective beam width greater than the joint gap. Seam tracking equipment ensures the beam is always aimed at the joint gap.

(e) Avoidance of fade out porosity defects which tend to form at the end of a weld when the beam is turned off is not yet fully resolved.

(f) It is sometimes difficult with non destructive testing such as ultrasonic examination to detect areas of only partial depth welding because the unfused areas of the mating faces can be in extremely close contact. Solutions to this problem are under investigation.

(g) The Welding Institute have a 60 kW non-vacuum EBW unit which has a maximum penetration of 40 mm in steel.

(h) EBW radiation in the form of the electron beam itself and secondary radiation excited in the weld parent metal would be compatible with penetrator assembly radiation from HLW as regards the need for shielding for operators. EBW vacuum chambers generally have lead cladding for this reason.

F.5 Laser Welding (LW) and Plasma Welding (PW)

(a) LW is an unlikely candidate at this stage as regards penetrator welding because of problems posed by thick wall welding. Current equipment has difficulties achieving 25mm deep welds in steel and LW is not being developed as a thick welding process.

(b) The concept of using LW equipment stationed outside and firing through a window in a radiation shielded cell to undertake penetrator assembly welding would be very attractive but for the weld penetration drawback.

(c) PW also has a limited penetration range. Current limits are about 10 or 12mm in steel and PW is not being developed for thick welding applications.

F.6 Friction Welding (FW) and Radial Welding (RW)

(a) FW is another strong candidate for penetrator welding. It is considered that a machine for welding 75mm wall thickness penetrators would be viable as it would only involve a relatively straightforward extrapolation of existing practices. However, FW of 200mm wall thickness penetrators cannot be assumed to be feasible at this stage as it is far beyond current experience of the technique.

(b) Approximate requirements for a machine to weld 75mm wall thickness penetrators would be in the region of 2000t thrust capacity and 1500 kW power rating. Rotation speed, temperature generated and welding time would be of the order of 50 rpm, 1350°C and 5 minutes.

(c) Currently the largest commercial machine in the UK has a 200 to 300t thrust capacity. An American company is reported to be building a 700t machine and actively studying the possibility of a 900t thrust rated version.

(d) UK designers prefer continuous drive as opposed to the stored energy, that is flywheel, machines used in the USA. Stored energy machines require greater thrust capacity than equivalent continuous drive units so it is not possible to compare UK and USA machines solely on the basis of their thrust rating.

(e) Although there are metallurgical changes in the region of friction welds, there is as yet no data to suggest that they give rise to preferential corrosion of the weld zone though very little research has been undertaken on the subject. FW has little difficulty meeting the requirements for less than 22 Rockwell C (248 Vickers) weld zone hardness often specified for oil pipeline and similar welds to limit susceptibility to sour gas (H_2S) attack.

(f) FW can join a wide range of similar and dissimilar metal combinations such as aluminium to steel. In particular, steels with up to about 2% carbon can be welded. However, higher carbon contents can cause a graphite lubricant effect which prevents generation of sufficient friction heat for welding.

(g) RW is a version of FW whereby the main components to be joined are held stationary whilst a solid ring of compatible material is rotated in the "V" joint between components. This generates heat and the ring is then compressed to complete the weld. It is possible that a RW technique could be suitable for penetrator welding, at least for 25mm wall thicknesses.

F.7 Diffusion Bonding (DB)

(a) DB of large area joints similar in size to those of the proposed penetrators has been reported by American workers.

(b) Typical present-day DB of steel entails heating to 1050°C in a 0.1 Torr vacuum, 10^{-4} Torr if induction heating is used, or an inert gas atmosphere. The components are then pressed together to develop approximately yield stresses at the mating faces until DB is complete, perhaps after 2 hours.

(c) Ra = 4 μm surface texture to BS 1134, that is very smooth mating surface preparation is used for current DB trials at the Welding Institute.

(d) It is possible to join a wide range of materials by DB. For example, pearlitic SG iron components with a 0.15mm nickel film between the mating faces have been successfully joined by DB.

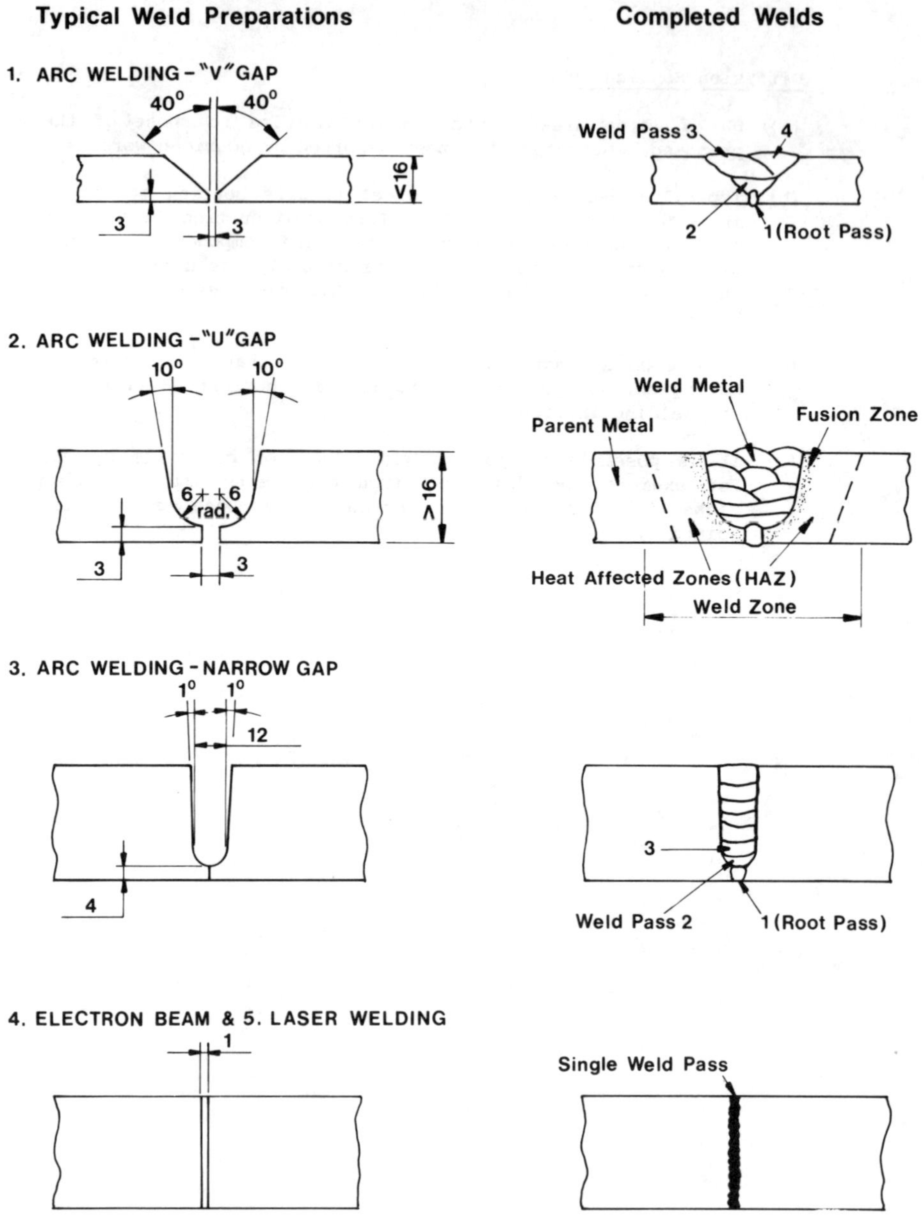

WELDING: FUSION PROCESSES

figure F1

Typical Weld Preparations | **Completed Welds**

1. FRICTION WELDING

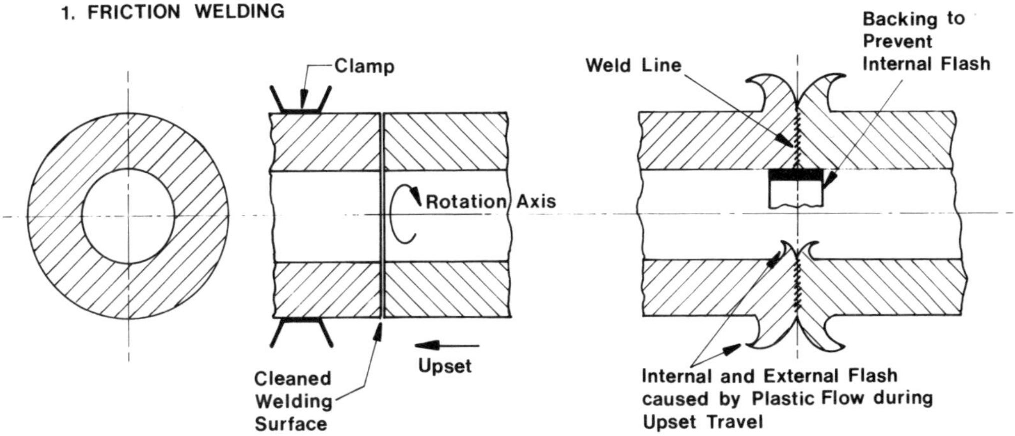

2. DIFFUSION BONDING

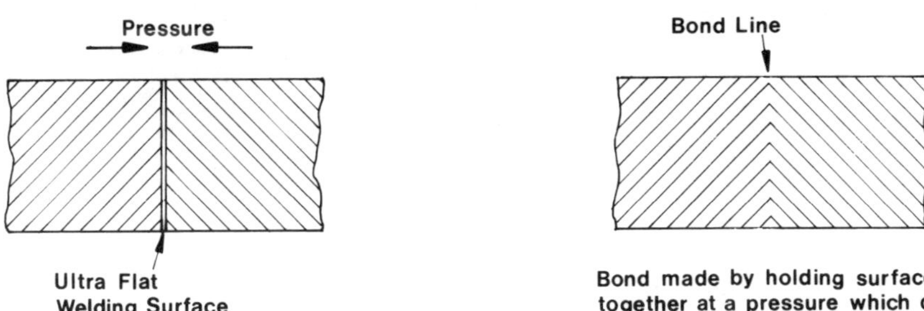

Bond made by holding surfaces together at a pressure which does not cause plastic flow and a temperature below the melting point (usually undertaken in a vacuum or inert gas atmosphere)

WELDING: SOLID PHASE PROCESSES figure **F2**

APPENDIX G

THERMAL EFFECT MODEL

APPENDIX G

THERMAL EFFECT MODEL

G.1 Introduction

This Section of the study considers the thermal effects associated with the heat output of the vitrified waste.

G.2 Heat Output and Data

The heat output from vitrified HLW reduces with time. For example, ten years after production, 0.5 tonnes of vitrified waste would produce about 1.5kW, whilst 60 years after discharge this output would have dropped to about 0.5kW.

The variation of waste heat output with time is indicated in Figure 3.

Table G.1

Ocean Sediment Thermal Properties

Property	Units	Value
Temperature at seafloor	°C	4
Temperature increase with depth	°C/km	54
Thermal conductivity	W/m/°K	0.8
Specific heat	J/kg/°K	3780
Diffusivity	m^2/sec	1.5×10^{-7}

G.3 Methodology

The rise in temperature adjacent to a waste container will depend on the initial heat output, the rate of decay in heat output, the length and diameter of the container, the spacing of the containers and the thermal properties of the sediment.

Hodgkinson and Bourke (Ref. 51) give the following expression for the approximate maximum temperature (T) adjacent to a cylindrical heat source based on the approximations of a line source and an infinite medium.

$$T = \frac{Q}{\pi kL}\left[A(L,r) - \frac{3L}{8(\pi g t_m)^{\frac{1}{2}}}\right]$$

where $A(L,r) = \frac{1}{2} \ln \left(L/2r + [1 + (L/2r)^2]^{\frac{1}{2}}\right)$

and t_m is the time to maximum temperature given by:-

$$t_m = \frac{1}{\pi g} [(\pi gL)/(8\lambda A(L,r))]^{2/3}$$

The above expression is reasonably accurate provided:

$$10\ r^2/g < t_m < 0.1/\lambda$$

Q	=	heat output of unit
L	=	length of unit
r	=	radial distance from centre line of source
λ	=	decay constant for the waste
k	=	thermal conductivity for formation
g	=	thermal diffusivity for formation

For the purposes of temperature calculations the following values were taken for a single waste container.

Heat output	Q	=	520 watts i.e. 50 year ex reactor
Length	L	=	1.6m
Radius	r	=	0.3m
Decay constant		=	7.33 E-10 s^{-1} corresponding to 30 year half life.

G.4 Limiting Temperatures

The limiting temperatures are controlled by the following factors:-

a) Recrystallisation of the borosilicate glass would occur at temperatures in excess of about 450°C and could cause an increase in the leach rate of radionuclides.

b) The maximum temperature for the surrounding sediments may need to be limited to avoid problems such as chemical alteration of clay minerals and increased pore-water convection.

Some proposed maximum overpack surface temperatures following emplacement are given below (Ref. 52) although it is unclear on what basis they have been chosen. They are therefore not intended to be design limitations and should be taken as provisional values only.

France	100°C
UK	100°C
USA	200 - 250°C

G.5 Effect of Spacing

The overall temperature distribution will be affected by the horizontal spacing of adjacent penetrators.

The calculation for an array of line heat sources indicates that there is a critical spacing beyond which the adjacent sources have negligible effect on the maximum temperature immediately adjacent a single source. Beyond this spacing, the only effect of adjacent sources is to prolong the time at which the temperature is close to a maximum.

For waste with a 30 year half-life, the critical spacing for a heat source representing a penetrator in a 2-dimensional array is approximately 15m in deep ocean sediments.

G.6 Peak Temperatures

For a length of heat source appropriate to penetrator the maximum temperature rise is calculated to be approximately 165°C.

The temperature at the seafloor is around 3°C and could be expected to increase at the rate of 54°C per 1000m. Thus the maximum sediment temperature for a reference penetrator embedded at 50m would be approximately 170°C.

One practical way to reduce this peak temperature would be to store the waste for a longer period than 50 years before disposal. If, for example, the waste were stored for 100 years before disposal then the heat output per canister would be approximately 180 watts and hence the maximum temperature would be approximately 63°C.